THIS IS
geography

JOHN WIDDOWSON

Although every effort has been made to ensure that website addresses are correct at time of going to press, Hodder Education cannot be held responsible for the content of any website mentioned in this book. It is sometimes possible to find a relocated web page by typing in the address of the home page for a website in the URL window of your browser.

Orders: please contact Bookpoint Ltd, 130 Milton Park, Abingdon, Oxon OX14 4SB. Telephone: (44) 01235 827720. Fax: (44) 01235 400454. Lines are open 9.00–5.00, Monday to Saturday, with a 24-hour message answering service. Visit our website at www.hoddereducation.co.uk.

© John Widdowson, 2006
Hodder Education
an Hachette UK Company
338 Euston Road
London NW1 3BH

Impression number   10 9 8 7 6 5 4
Year   2012 2011 2010 2009

All rights reserved. Apart from any use permitted under UK copyright law, no part of this publication may be reproduced or transmitted in any form or by any means, electronic or mechanical, including photocopying and recording, or held within any information storage and retrieval system, without permission in writing from the publisher or under licence from the Copyright Licensing Agency Limited. Further details of such licences (for reprographic reproduction) may be obtained from the Copyright Licensing Agency Limited, Saffron House, 6–10 Kirby Street, London EC1N 8TS.

Layouts by Ama...
Artwork by Art ...k, Oxford Designers and Illustrators Ltd, ...

Typeset in 12/14...
Printed and bou...

A catalogue record for this title is available from the British Library

ISBN: 9 780340 912195

Teacher's Resource Book including CD-ROM
ISBN: 9 780340 907443

eLearning Activities CD-ROM
ISBN: 9 780340 907474

Contents

Key features of *This is Geography*		2
What is geography?		4

Unit	Your final task	
1. Your place ... and mine! Is there a place for everyone in your school?	Draw a plan for your school, to make it more accessible	8
2. My spaces How can we use maps to show our lives?	Produce a wall chart showing your life and how it is connected with the spaces around you	20
3. Survivor! Could geography help you to survive an island adventure?	Choose ten items that could help you survive on a deserted island	34
4. City – past, present, future Why do people choose to live in cities?	Identify problems in 'Metropolis' and plan improvements	52
5. Shop until you drop! How is the way that we shop changing?	Write a children's story about shopping	70
6. Flood disaster How could we be better prepared next time?	Design a flood warning poster	84
7. What a load of rubbish! What should we do with all our waste?	Write a letter to the local council about how to deal with waste	100
8. Look again at the United Kingdom What would a newcomer to the UK want to know?	Make a PowerPoint presentation about the UK for newcomers	114

Key concept table	132
Reference map of the UK	133
Ordnance Survey 1:50,000 map symbols	134
Glossary	136
Index	138
Acknowledgements	140
Where next?	141

Key features of This is Geography

Before you start *This is Geography*, here is a quick guide to help you find your way around. Book 1 is split into eight units, covering eight enquiries. In each one you will find the following features:

The opening spread

The unit title

Key concepts covered in this unit.

The big enquiry question – the main question you will focus on through the unit.

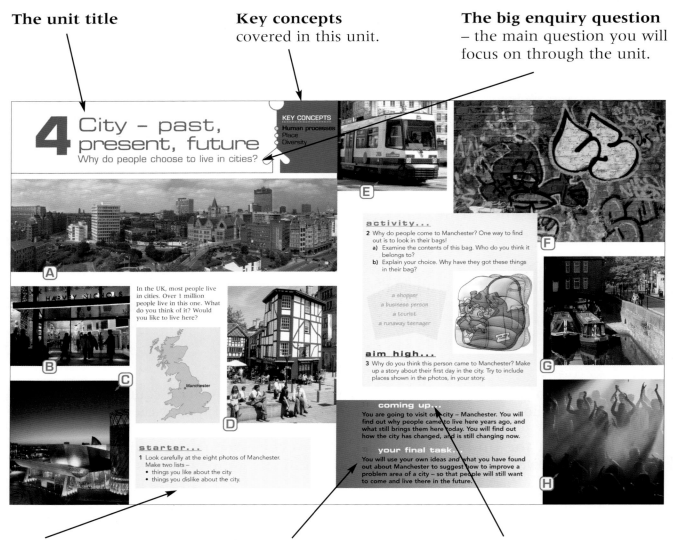

Starter – a fun activity to get you started and to make you think.

Your final task – what you will be doing at the end of the enquiry to bring all your work together.

Coming up – tells you what you are going to do through the rest of the unit.

KEY FEATURES OF THIS IS GEOGRAPHY

Through the unit

Key words – in SMALL CAPITALS. They are explained in the Glossary on pages 135–136.

Activity – tasks that will help you to build on your enquiry, step by step.

The enquiry question – repeated on every spread so you won't forget it!

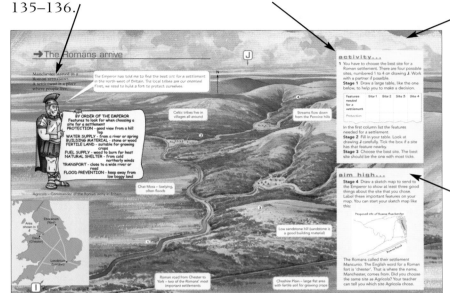

Step-by-step explanations of important geographical skills. All are animated on the CD-ROM.

Aim high – a challenge, and not just for the clever ones! It is a task that will help you to take your geography that little bit further.

The final spread

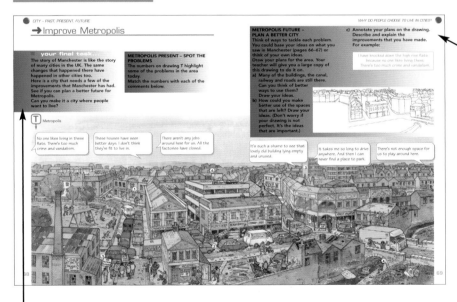

The enquiry question – again! You should know it by now.

Your final task – this is what all your work has been leading to. It is your chance to show what you know and what you can do. You may be asked to:
- draw a plan
- make a wall display
- write a story
- design a poster
- make a PowerPoint presentation.

… And for every activity you will find worksheets and eLearning activities to help you.

What is geography?
Asking questions to make sense of the world

Who...?
Where...?
How...?
What...?
Why...?

A

B

discuss...
1 What do you think geography is about? Discuss this with a partner. Try to come up with at least five ideas.

ASKING QUESTIONS TO MAKE SENSE OF THE WORLD

C

D

activity...

2 Now look at photos **A** to **D**.
   a) Which photo is most like your idea of geography? Why?
   b) Which photo do you find most interesting? Why?

3 You will be given a copy of the photo you chose in 2 b).
   a) Stick it in your book.
   b) Write at least five interesting questions around the photo – things you would like to know the answer to. They could begin with:
   Where ...?
   Who ...?
   What ...?
   Why ...?
   How ...?

WHAT IS GEOGRAPHY?

➜ What will I learn in geography?

In geography you will study …

Places

You will start with places you know – like your school and the local area. You will then go on to study more places in the United Kingdom and places in other parts of the world.

Themes

Geography helps you to understand the world. **Physical geography** is about our planet. It includes themes like rivers, weather and volcanoes. **Human geography** is about people. It includes themes like cities, population and work.

Issues

Geography explains the issues you read about in the news. It can even help find solutions to problems such as:
- what to do with our waste
- how to make cities better places to live
- the consequences of global warming.

Underlying everything you study in geography are **seven key concepts**, or big ideas. Think of them like the parts in your body. Normally, you can't see them, but you know they are important.

ASKING QUESTIONS TO MAKE SENSE OF THE WORLD

discuss...

1 Read this list of things you could study in geography. Sort them into three groups – **places**, **themes** and **issues**. You can use the glossary at the back of the book to help you.

> Refugees
> Colorado River, USA
> Population
> Coasts
> Drought
> London
> Landslides
> Kosovo, Eastern Europe
> Rivers
> Regeneration
> Cities
> Scarborough, Yorkshire

2 Look again at photos **A–D**. Which place, theme or issue might you be studying in each photo?
  a) Match each photo with one **place**, one **theme**, and one **issue**.
  b) Choose one photo. What does it tell you about that **place**, that **theme**, and that **issue**?

aim high...

3 Think about how you could link the seven key concepts to the photos.
  a) Choose a photo. It could be the same photo you chose in 2b, or a different one. Stick a copy of the photo in your book.
  b) How does the photo link to any of the key concepts? Write your ideas around the photo. For example, if you chose photo **C**, you could write:

> The photo shows a place. It is the Colorado River in the USA.
> The river is flowing through a vast space. It looks like a desert.
> Two tiny boats show the huge scale of the river.
> The only visible interaction between people and the environment is the boats on the river.
> The river has carved a deep valley. This process would take a long time.

1 Your place... and mine!

Is there a place for everyone in your school?

KEY CONCEPTS
- **Place**
- Scale
- Diversity

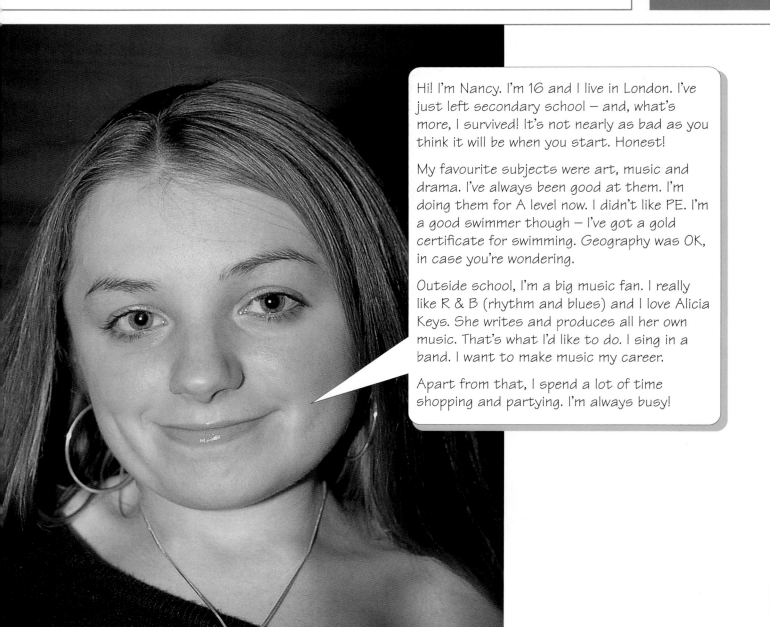

Hi! I'm Nancy. I'm 16 and I live in London. I've just left secondary school – and, what's more, I survived! It's not nearly as bad as you think it will be when you start. Honest!

My favourite subjects were art, music and drama. I've always been good at them. I'm doing them for A level now. I didn't like PE. I'm a good swimmer though – I've got a gold certificate for swimming. Geography was OK, in case you're wondering.

Outside school, I'm a big music fan. I really like R & B (rhythm and blues) and I love Alicia Keys. She writes and produces all her own music. That's what I'd like to do. I sing in a band. I want to make music my career.

Apart from that, I spend a lot of time shopping and partying. I'm always busy!

IS THERE A PLACE FOR EVERYONE IN YOUR SCHOOL?

starter...

1 Read what Nancy says or listen to it.
2 Discuss what it tells you so far about Nancy. Record it on a spider diagram like this.

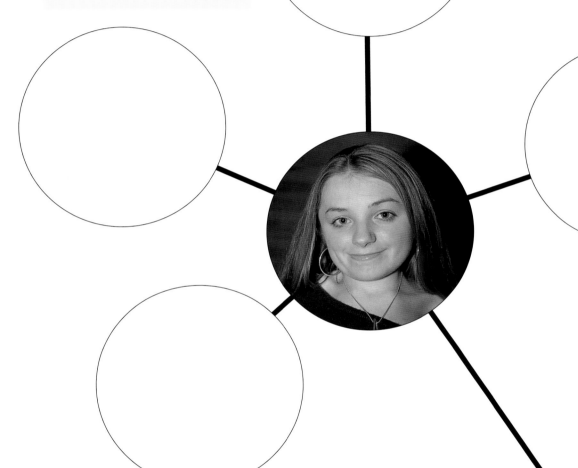

through the unit...

You might be asking yourself what Nancy's story has got to do with geography.
Geography is not just about places. It's about people too. Everyone has their own personal geography – the places where they live, visit and dream of going to. You will think more about *your* personal geography in Unit 2. Keep asking yourself the question: 'What has Nancy's story got to do with geography?' At the end of this unit you will try to answer this.

YOUR PLACE... AND MINE!

➔ Find out more about Nancy

There's something else you should know about me. I have brittle bone disease. It's not catching! It's genetic — something I was born with. Basically, it means that my bones are really fragile. They can break if I just bump into something. I have to be extra careful in school. I avoid busy corridors and rough sports.

I've been in hospital with broken bones more times than I can count. I bet you think it's really cool to get so much time off school. Believe me — it's not! It really messes up your education. I was lucky that my mum helped me a lot at home, so I passed my GCSEs.

At school I was usually in a wheelchair. I can walk a bit, but it gets really tiring.

One other thing: I'm shorter than most people because my bones don't grow. I'm only four feet six inches.

discuss a...

1. Read what Nancy says. Add this information to your spider diagram.
2. How did your view of Nancy change when you read about her disability?
3. How do you think Nancy would fit in at your school?

IS THERE A PLACE FOR EVERYONE IN YOUR SCHOOL?

discuss b...

4 Work in pairs. Look at photos **A** to **D**.
  a) Imagine you are in a wheelchair. What problems would you face in each photo?
  b) Think of a solution to the problems in each photo.

coming up...

In this unit you have to put yourself in Nancy's shoes. What would it be like in **your** school if you were in a wheelchair? Do the buildings need to change? Do attitudes need to change?

your final task...

You will make a plan for your school to make it more accessible.

11

YOUR PLACE... AND MINE!

→ Wheelchairs welcome?

Finding a secondary school was a problem. Only one local school had access for wheelchairs. I went with my parents to look at it. On the way round we saw a fight. We thought it was not the best place for someone with brittle bones!

I was offered a place in a special school for disabled kids. My mum and dad said, 'No way!' They wanted me to go to school with ordinary kids. I might be in a wheelchair but my brain's OK. They had to go to a tribunal (like a court) to fight my case. Eventually, we won and I went to a secondary school in another area.

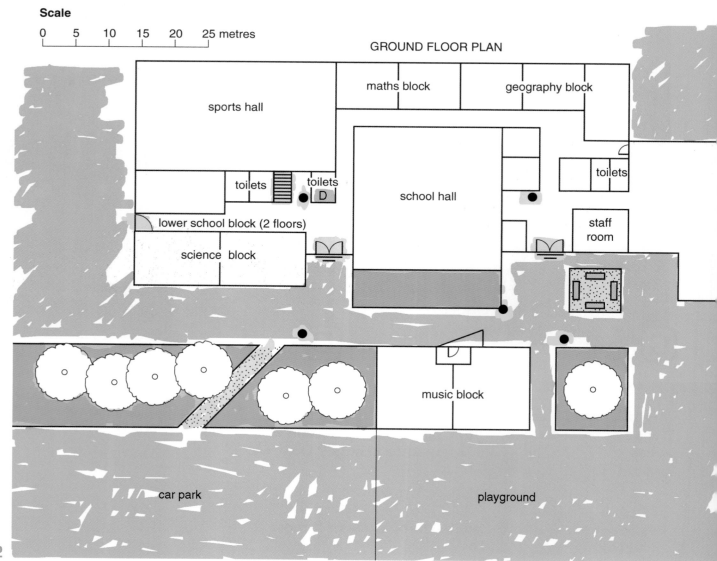

ACCESS is how you get to a place. Somewhere that is accessible is easy to get to.

activity...

1. Look at plan **E**. It is an access survey of a secondary school.
   a) Identify all the things that make it easy or difficult to get around the school in a wheelchair. Write two lists. Think of anything you could add to your lists that is not on the plan.
   b) **E** is a ground floor plan. Why couldn't pupils in wheelchairs get to the other floors? How could the problem be solved?
2. You are going to do an access survey of your school. Depending on the size of your school, the class might need to divide into groups. Each group will survey part of the school. Your teacher will give you instructions.
   Work in a group.
   You will need:
   - a wheelchair (it is possible to do the survey without a wheelchair, but you will need to think carefully about wheelchair access)
   - a plan of the school and a clipboard
   - two different-coloured pencils.
   a) One person in the group should sit in the wheelchair. The rest of the group will push it around. As you go around the school, notice things that are wheelchair-friendly or wheelchair-unfriendly. Record them in the correct place on your plan, using shading or symbols (plan **E** shows you how to do this). Include the type of surface and other features. Use two colours: one for things that are wheelchair-friendly and one for things that are wheelchair-unfriendly.
   b) Back in the classroom, share your findings with the rest of the class. Complete an access plan for the whole school. Draw symbols and a key, similar to plan **E**. You can make up your own symbols for other features you found.

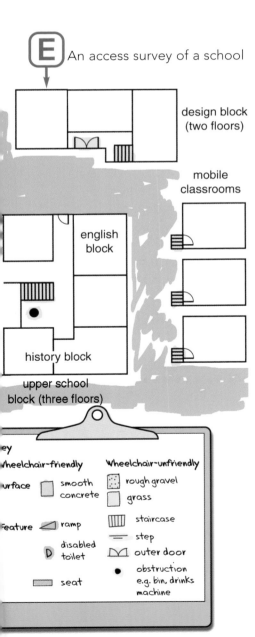

E An access survey of a school

aim high...

3. Write a short report on accessibility for your school. It could have three paragraphs.
   - In what ways is the school accessible for wheelchairs?
   - In what ways is it inaccessible?
   - How could access be improved?

YOUR PLACE... AND MINE!

→ Access all classrooms

"Even though I went to a school that was meant to be accessible for wheelchairs, there were problems. For a start, it was difficult to fit a wheelchair into some classrooms."

how to..... Draw a scale plan

A PLAN is like a map. It is what you would see if you look down from above. Here is how you draw a plan of a table.

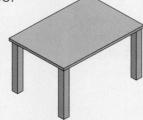

Normal view of a table **Plan view of a table**

If you want to be accurate, you draw a SCALE PLAN. The SCALE on the plan tells you the size in real life. To draw a scale plan of your table or desk:

1 Measure the sides of the table with a ruler.
2 Work out how much you have to reduce your table to make it fit on a page in your book. For example, if the long side of your table is 120 cm you could reduce it ten times to make it 12 cm. You then divide all the other measurements by 10 as well to match.
Draw a plan of the table in your book with these measurements.
3 Add the scale next to your drawing. For example, the scale 1:10 means that every 1 cm on your plan would be 10 cm in real life. Another way to show your scale is to draw a scale line. Mark off the line with the real-life measurements.

```
0                    50                  100 cm
```

activity a...

1 Here are some drawings of things that you might find in a classroom.
   a) Copy the drawings into your book or draw three things in your own classroom.
   b) Now, draw a plan of each one. If you are able to measure it, draw a scale plan. Use the scale 1:10, or choose your own scale.

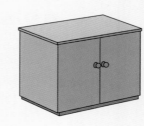

activity b ...

2 Look at plan **F**. Notice the scale on the plan.
   a) Cut out a piece of paper exactly the same size as this. This is a scale plan of a wheelchair! The reason that it looks so small is it has a scale of 1:50 (the same scale as plan **F**).
   b) How would a wheelchair fit into the classroom? To find out, place your piece of paper on the plan. Try coming through the door and then moving to a desk so that you can sit behind it. How difficult was it?

3 Change the layout of the classroom to make it more accessible for wheelchairs.
   a) Draw the outline of the classroom exactly the same size as plan **F**. Give it the same scale.
   b) Cut out 15 tables the same size as the tables on plan **F** (there should be space for up to 30 pupils). Do the same for the other furniture. Your teacher may give you a sheet to make it quicker.
   c) Place the furniture on the plan. Rearrange the furniture to make the classroom more accessible. For example, you could try putting tables in groups or rows.
   d) When you think you have the best layout, stick the furniture in place on the plan.

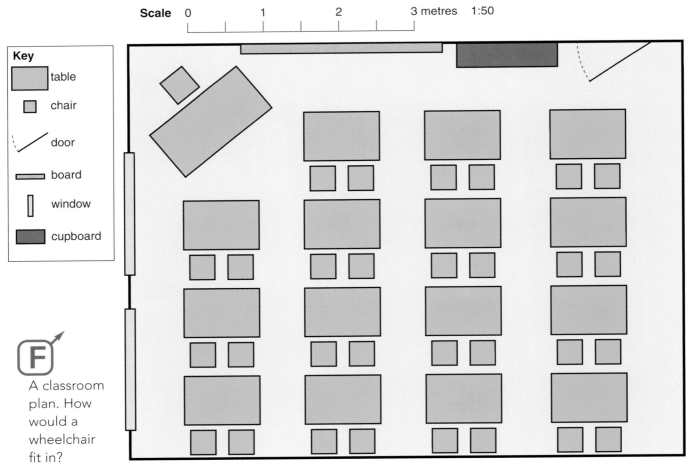

F A classroom plan. How would a wheelchair fit in?

YOUR PLACE ... AND MINE!

→ It's about attitudes too

Access is not just about buildings. It's about people's attitudes too. I found this out at school. Getting around the building was the easy bit. The main problem was people. Some of them thought that they knew what was best for me. None of them bothered to ask me!

activity...

1 Look at the cartoons in **G**. Don't forget – these were real situations!
In each situation:
a) Imagine how Nancy felt. Write down what you think her feelings were.
b) Compare what you wrote with what Nancy herself says here. Read what she says in the boxes.

> I really wanted to go on school trips. I'm sure they couldn't be that dangerous. I expect the head was worried that something might go wrong. But I was prepared to take a risk and it is my body.

> I didn't want a support worker. It made me different from other pupils. They didn't talk to me when an adult was there. If I needed help I'd rather ask another pupil.

> I was scared of playing basketball. I was worried that the ball would hit me and break a bone. The teacher didn't ask me why I didn't want to play.

c) Suggest what the school should do to include disabled pupils more.

PE lessons

> You ARE going to play basketball young lady. No excuses! Some of the best basketball players are in wheelchairs. There's nothing to stop you playing.

> But...

G Attitude problems! The cartoons show real situations that Nancy found herself in at school

IS THERE A PLACE FOR EVERYONE IN YOUR SCHOOL?

discuss...

2 Talk with a partner.
What are the benefits if all pupils attend the same school?
   a) for disabled pupils
   b) for other pupils in the school.
Make a list of the ideas you think of.

aim high...

3 Some people argue the opposite. They say that it is not sensible for able-bodied and disabled pupils to go to the same school. Write a paragraph to explain whether you agree or disagree with this view. Think about what you have learnt so far in this unit. Give evidence to support your view.

The support worker

School trips

YOUR PLACE... AND MINE!

→ Count me in!

For a long time disabled pupils were SEGREGATED. They were taught in separate schools. This did not prepare them for the real world. Being kept apart also made people think that disabled children were different.

Now the government wants schools to be INCLUSIVE. That means disabled and able-bodied pupils will be taught together. Is your school ready? If not you can help it get ready.

> I just want to be treated like everybody else. The last thing I'd want is to be stuck away in a special school. Ordinary schools need to change so that there is a place for people like me.

Classrooms have wide entrances and no doors (so wheelchairs can easily get in and out).

The school has wide corridors and is all on one level (to make it easy to move around).

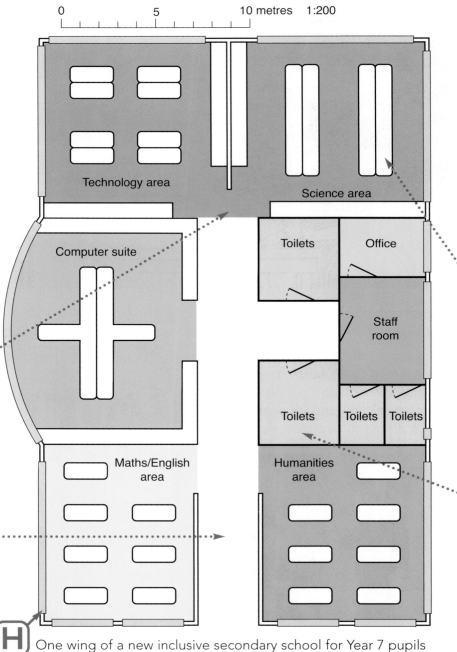

One wing of a new inclusive secondary school for Year 7 pupils

IS THERE A PLACE FOR EVERYONE IN YOUR SCHOOL?

your final task...

You are going to produce a plan to make your school more inclusive. It could be a plan for your classroom, the geography area or one particular problem area of your school that your access survey from page 13 identified. If you are really ambitious you could do a whole school.

1 Start with a plan of what it is like now (you should already have this from earlier in the unit – or your teacher will give you one to work on).
   a) **Think** about the changes you could make. The plan of the school opposite might give you some ideas.
      - You could change the layout of the furnishings – for example, reduce the number of tables in a classroom.
      - You could include alterations to the buildings – for example, replacing steps with a ramp, or knocking down some walls to make bigger spaces.
   b) Add labels or stickies to show the problems and possible solutions.
2 Now **draw** a plan to show your changes.
   a) Use a different colour to highlight your changes.
   b) Label your plan to explain the changes you have made.

aim high...

3 Remember, it's not just about buildings. It is also about attitudes.
Add some bullet points to your plan to show how your redesign would help everyone in the school develop a positive attitude to disabled pupils.

through the unit

Now that you have worked through this unit think again about the following question: 'What has Nancy's story got to do with geography?'. Can you answer this now?

Furniture in the classrooms is widely spaced (to allow room for wheelchairs).

Toilets are close to classrooms (so pupils don't have far to go).

I This is a screenshot from 'Schools of the Future', a video produced by the Department for Education and Skills. It shows what an inclusive school could look like in the future.

2 My spaces

How can we use maps to show our lives?

KEY CONCEPT
- **Scale**
- Place
- Space

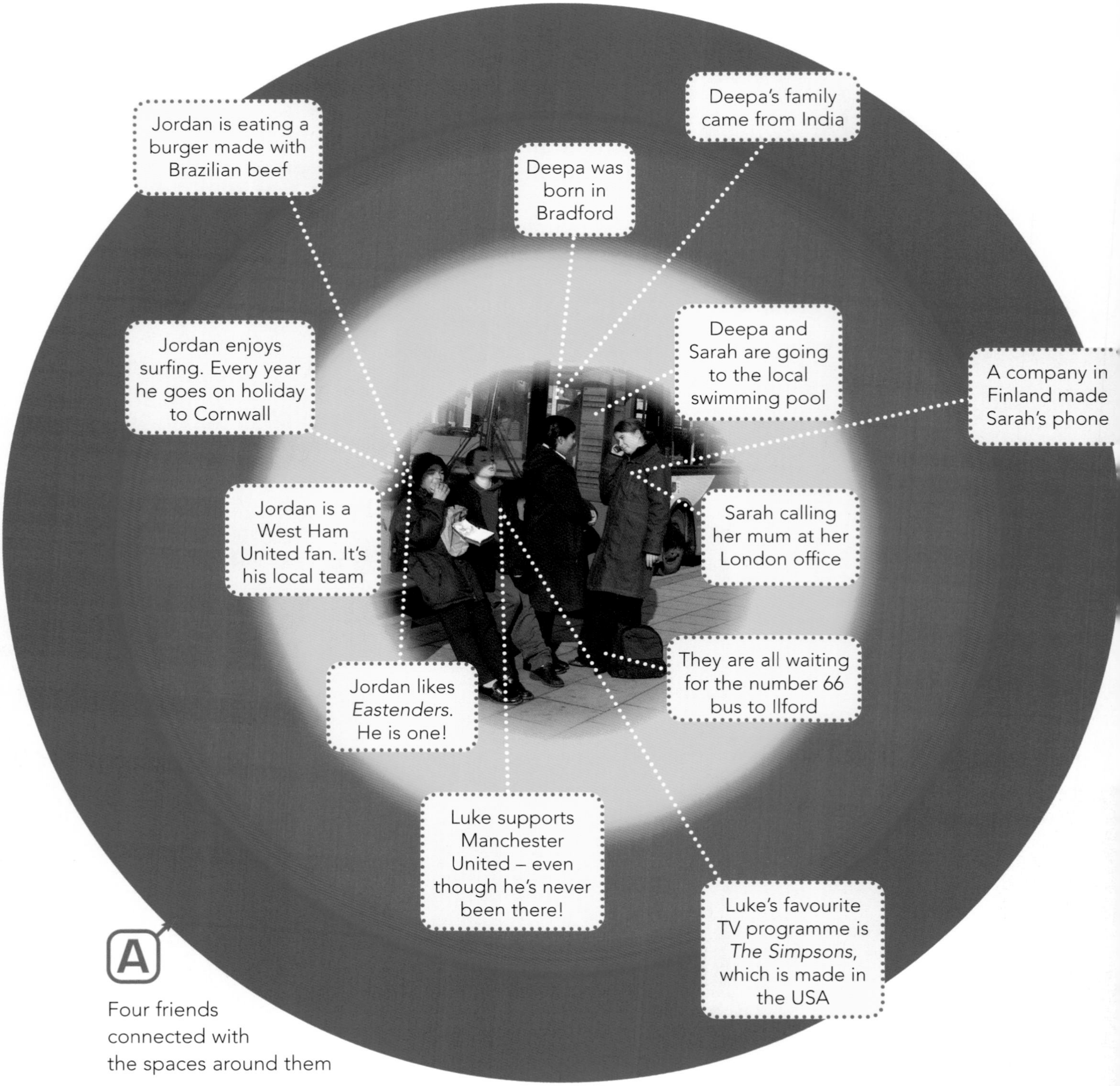

- Jordan is eating a burger made with Brazilian beef
- Deepa was born in Bradford
- Deepa's family came from India
- Jordan enjoys surfing. Every year he goes on holiday to Cornwall
- Deepa and Sarah are going to the local swimming pool
- A company in Finland made Sarah's phone
- Jordan is a West Ham United fan. It's his local team
- Sarah calling her mum at her London office
- Jordan likes *Eastenders*. He is one!
- They are all waiting for the number 66 bus to Ilford
- Luke supports Manchester United – even though he's never been there!
- Luke's favourite TV programme is *The Simpsons*, which is made in the USA

A Four friends connected with the spaces around them

HOW CAN WE USE MAPS TO SHOW OUR LIVES?

We all live in space! Your local area is a small space. Many of us live in the United Kingdom, which is a much bigger space. The biggest space of all is the world. That's where we all live. All these spaces can be shown on maps.

The friends in photo **A** go to the same school in London. They have **local**, **national** and **global** connections. So do you!

starter...

Look at photo **A**.
1 Make a list of all the places that the friends are connected to.
2 a) Draw a large copy of this diagram.

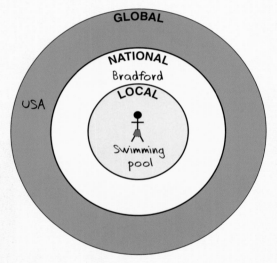

b) Write each place in the correct space on your diagram. One of each is done for you.

my spaces...

3 Now write down places you are connected to. Here are some questions to get you started:
 - Where do you, your parents, and your grandparents come from?
 - Where do you go on holiday?
 - Where do you spend your leisure time?
 - Where do your food and clothes come from? Find out by looking at labels.
 - Where do you get phone calls, texts or e-mails from?

4 Draw a large circular diagram with three spaces like the one for activity 2. To show *your* connections you can:
 - put the name of the place in one of the spaces and explain the connection, for example:

 Manchester – the football team I support.

 - or, attempt the Aim High challenge in activity 5.

aim high...

5 How good is your sense of direction? Before you add any places to your diagram, draw a compass like this at the centre. Write each place you are connected with, and mark the direction you would have to go to get there in real life. For example, the USA is to the west.

coming up...

In this unit you will use maps to explore your spaces at a range of scales – from your local area to the whole world. You will learn some new geographical skills to help you.

through the unit

Whenever you do a My Spaces activity you will produce a map that you need to save for your final task.

■ your final task...

At the end of the unit you will use all your maps to produce a wall display in your classroom to show your life, and how it is connected with the spaces around you.

MY SPACES

→ My school

Sarah, Luke, Jordan and Deepa go to Wanstead High School in east London. The school is somewhere in photo **B**. This is an AERIAL PHOTO or bird's eye view. It shows you things that you might not see if you were on the ground.

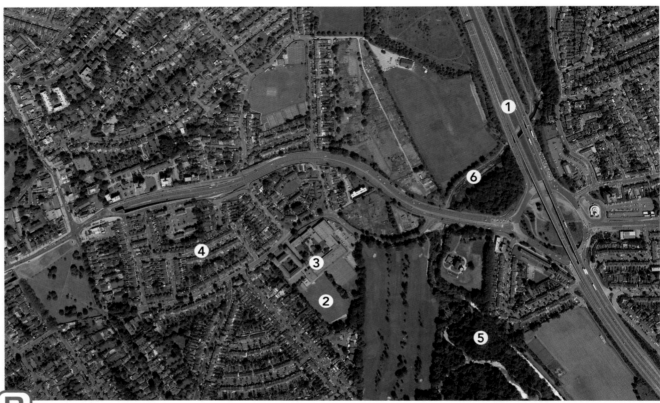

B Aerial view of the area around Wanstead High School

activity a...

1 Look carefully at photo **B**.
   a) Find each of these features in the photo: the school, houses, trees, the school playing field, a main road, and a river.
   b) Match them with the numbers on the photo.
2 Look carefully at map **C**.
   a) Name the main road that goes from east to west.
   b) Name the road that the school is on.
3 Compare photo **B** and map **C**.
   a) What can you see in the photo that you cannot see on the map? Mention three things.
   b) What can you see on the map that you cannot see in the photo? Mention three things.
   c) Which do you think gives the best idea of what the place is like – the photo or the map? Explain why.

HOW CAN WE USE MAPS TO SHOW OUR LIVES?

Maps also give us a bird's eye view of the world. Map **C** shows the same area you can see in photo **B**. Which one gives you the best idea of what the place is like?

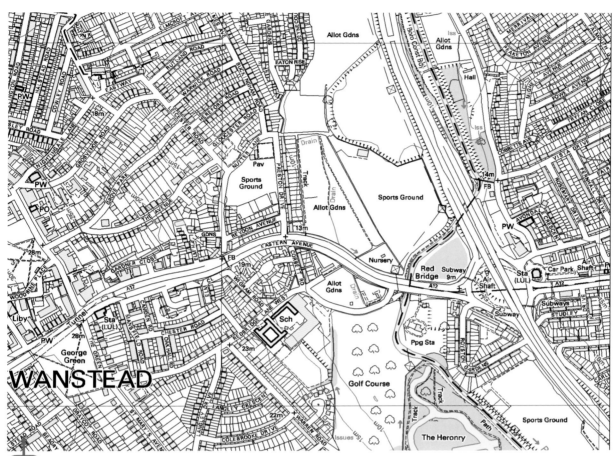

C Map of the area around Wanstead High School.

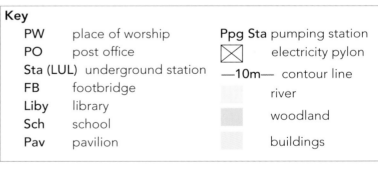

activity b...

4 You can find the same photo and a map on the Multimap website, at www.multimap.com

  a) This is what you need to do:
  - Type in the postcode 'E11' in the 'Search' box. Click 'Find' and a map will appear.
  - Click on 'Wanstead' to zoom in. You should see a map like map **C**.
  - Finally, click on 'Aerial' above the map. It will turn into a photo, like photo **B**.

  b) Using this method, find an aerial photo and map of the area around your school. Start with the postcode. What features can you find around your school?

my spaces...

Choose the best way to show the area around your school. You could choose an aerial photo or a map from the Multimap website.

Keep your photo or map to use in the classroom display at the end of the unit.

MY SPACES

→ My area

Sarah, Luke, Jordan and Deepa all came from different primary schools in and around Wanstead. You can find the schools in map **D**. They became friends when they met at Wanstead High School. It is in square C3 on map **D**.

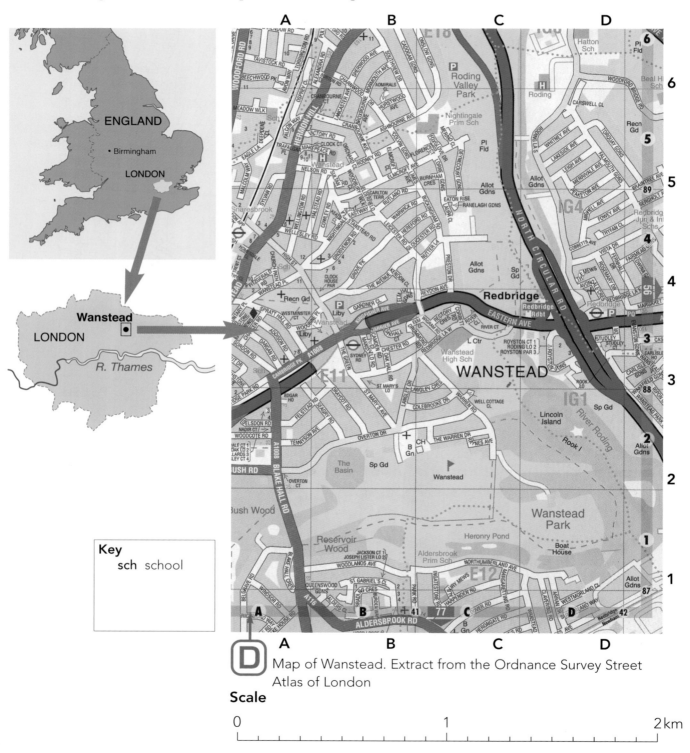

Key
sch school

Map **D** Map of Wanstead. Extract from the Ordnance Survey Street Atlas of London

Scale
0 — 1 — 2km

HOW CAN WE USE MAPS TO SHOW OUR LIVES?

activity...

1. a) Find the primary school that each pupil went to. Use the grid to help you.
   Sarah went to Nightingale Primary School in square C5.
   Luke went to Snaresbrook Primary School in square A5.
   Jordan went to Wanstead Church of England Primary School in square A4.
   Deepa went to Aldersbrook Primary School in square C1.
   b) Measure the straight line distance from each primary school to Wanstead High School. Use the scale line below the map to do this. (See the How to... box.)
   c) Which primary school is furthest from the High School?

2. a) Read Sarah's description of the route from her primary school to Wanstead High School. As you read, follow the route on the map.

   I turn left from Nightingale School onto Ashbourne Avenue. I turn left again onto Colvin Gardens. I walk to the end of the road and cross over. I go straight along Limes Avenue and Buckingham Road to Nutter Lane. I turn right and walk to Eastern Avenue. I cross the bridge over Eastern Avenue and walk along Wigram Road to Wanstead High School.

   b) Choose another pupil. Describe the route from their primary school to Wanstead High School. Use Sarah's description as your model.

aim high...

3. Measure the distance that Sarah walks on her journey to Wanstead High School. Clue: measure each section of the journey and add them together.

how to... Measure distance on a map using the SCALE

1 Place a strip of paper between two points on the map.

2 Mark the two points on the paper.
3 Transfer the strip to the scale line. Place the first point on 0. Read off the distance to the second point on the scale line.

my spaces...

You will need a local street map that shows your secondary school.
a) Find your secondary school on your map.
b) Find another place on your map that you know well. It could be your old primary school, your house, the station where you catch the train home, or somewhere else. Work out a route from your secondary school to the place you choose. Write a paragraph to describe your chosen route.
c) Now read out your description to your partner so that they can follow it on the map. They have to find the place it leads to.
d) Finally swap roles. Follow your partner's route to find their place.

Draw your route on a map. Keep it to use in the classroom display at the end of the unit.

25

MY SPACES

→ My holiday

Every year Jordan goes on holiday to Cornwall. His family rents a caravan near the town of Newquay. This area has some of the best surfing beaches in Britain. Jordan loves surfing. Sometimes, he wishes that he lived here!

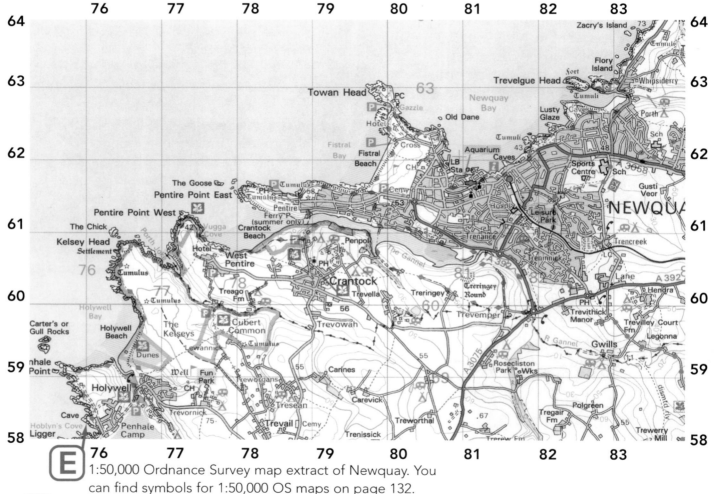

E 1:50,000 Ordnance Survey map extract of Newquay. You can find symbols for 1:50,000 OS maps on page 132.

F

activity...

1 Look at map **E**. The caravan site where Jordan stays is in square 78 58.
   a) Find the caravan site on the map. What is the name of the village?
   b) Nearby are lots of surfing beaches. Give the four-figure grid references for:
      i) Newquay Bay
      ii) Holywell Bay
      iii) Fistral Bay

HOW CAN WE USE MAPS TO SHOW OUR LIVES?

how to... Use grid references

Each square on a map has its own GRID REFERENCE. The important rule to remember to find a grid reference is – GO ALONG THE CORRIDOR, THEN UP THE STAIRS – *along* then up. For example, to find grid reference 81 61:
1 First, read across the bottom of the map to the line 81 (along the corridor).
2 Now, read up the side of the map to the line 61 (up the stairs).
3 You have found square 81 61. This is a **four-figure grid reference**.

But if you need to find a specific building on a map you need to be more accurate. Newquay Station is at 816 617 on the map.
1 Imagine the sides of each grid square are divided into ten equal parts. Read across square 81 to the imaginary line 6. This is 816.
2 Now, read up square 61 to the imaginary line 7. This is 617.
3 You have found the point 816 617. This is a **six-figure grid reference**.

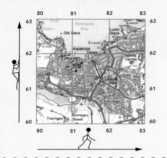

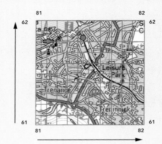

activity...

2 When he's not at the beach, Jordan likes to walk along the coast path (labelled SWC Path) on the map. The views from the headlands are awesome.
   a) Follow the path on the map from Penhale Point (756 592) to Trevelgue Head (823 621).
   b) If he were at these grid references, which headland would he be standing on?
      i) 764 608
      ii) 780 616
      iii) 797 630
3 Apart from its beaches, Cornwall has lots of other things to offer holidaymakers. Give the six-figure grid references for:
   a) the caves in Newquay
   b) the hotel at West Pentire
   c) the fun park at Trevornick.

my spaces...

You will need a 1:50,000 map of an area where you have been on holiday in Britain. Your parents might have one. If not, imagine a holiday in Cornwall, using map **E**.

a) Think of all the activities you did and the places you went to on your holiday. Find their location on the map. (If you stayed in one place, find the places you could have gone to!)
b) Make a table, listing the activities you could do on holiday in this area. Don't forget the obvious ones like swimming, walking, surfing, and so on.
c) Give six-figure grid references for the places where you could do these activities on the map.

Keep your table and map to use in the classroom display at the end of the unit.

MY SPACES

→ My country

Remember Luke, the Manchester United supporter. He's never been to Manchester! But he has been to the places shown in the three photos here. You can find them on ATLAS map **J** of the British Isles.

activity...

1. a) Name each of the places shown in the photos **G–I**. Can you find them on Map **J**? Use the clues to help you.
   b) Make up a geographical clue for the city where you live.

my spaces...

2. a) Look at Map **J**. List any cities on the map that you have visited.
   b) On a blank map of the UK show the cities that you have visited.
3. a) Think about some other places that you know in the UK. For example,
     - places that you go to on holiday, or for day trips
     - places where you have friends or relatives.
     Make a list.
   b) Locate these places in the UK with the help of the map on page 134. Add them to your map of the UK. Use a different colour for each category.
   Keep your map to use in the classroom display at the end of the unit.

aim high...

4. Look carefully at Map **J**.
   a) Can you think of a more accurate way than the clues used here, of giving the position of a city on the map?
      (Hint: use the grid lines!)
   b) Choose a city. Try this method out on a partner. Can they find the city?

G Clue 1: I have recently visited my sister. She is studying Geography at one of the most famous universities in the world.

Clue 2: West of London
On the River Thames

H Clue 1: Sometimes we go back to my dad's home city. They've built a new bridge over the river since we were last there.

Clue 2: In the north-east of England
Close to the North Sea

I Clue 1: This is where we go on holiday. I love eating rock and walking down the pier!

Clue 2: On the south coast
One hour from London

HOW CAN WE USE MAPS TO SHOW OUR LIVES?

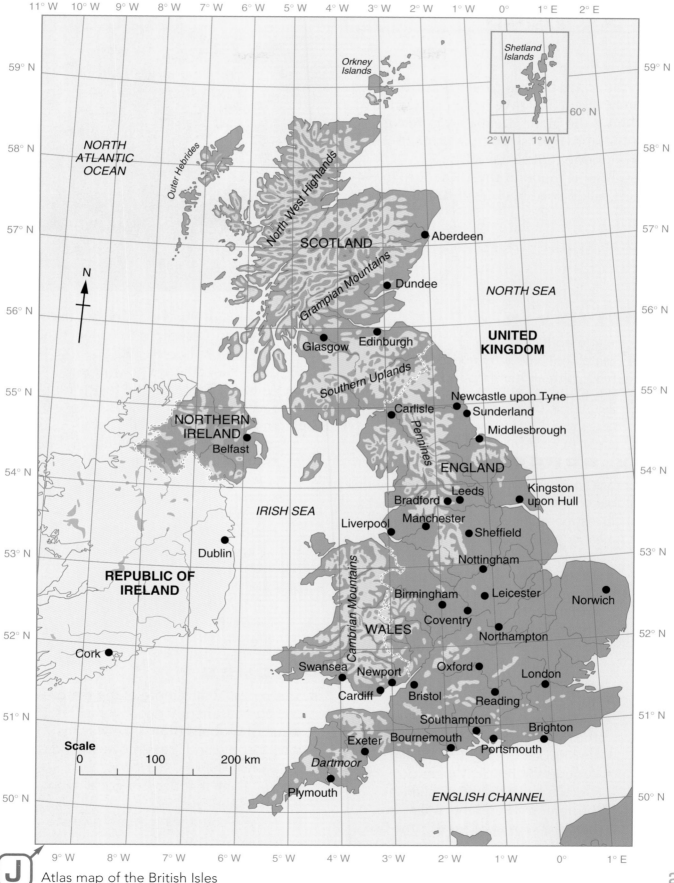

J Atlas map of the British Isles

MY SPACES

→ My world

Sometimes, it's easy to forget that the world is one of our spaces. We have connections all around the globe. Deepa's family come from India. A Finnish company made Sarah's phone. Luke's favourite TV programme is made in the USA. Jordan eats burgers made with beef from Brazil.

activity a...

1 Look at map **K**.
   a) Match each country numbered on the map, with an item around the map. Try this first without an atlas.
   b) Check your answers on a world map in the atlas.

my spaces...

2 Create your own world map to show your global connections.
   a) Investigate things in your home. Make a list of all the countries they come from. Here are some places to look:
      • The kitchen. Where does your food come from? Read the labels on tins and packets. Even better, go shopping! Read the information about the products you buy – especially fresh fruit, vegetables and meat.
      • Your bedroom. Where do your clothes and shoes come from? Read the 'Made in…' labels hidden on the inside.
      • Around the house. Where do electrical goods and appliances come from? Read the 'Made in…' labels on the appliance or the box.
   b) Find all the countries on your list on a world map in the atlas.
   c) On a blank world map, colour the countries. Draw lines to connect them with the UK, like map **K**.
   d) Label the countries on your map. Write the names of the things from each country on the lines.

3 a) Think about your other global connections. For example:
      • countries where you have been to on holiday
      • countries where you have friends or relatives
      • other ideas of your own. For example, where your favourite film was made.
   b) Find the countries in the atlas. Add them to your world map. You could use a different colour for each category.

discuss

4 These days, many more of the things we buy come from other countries. This is a result of **globalisation** – the way that jobs, people and ideas move around the world.
   Do you think that this is a good thing:
   a) for us?
   b) for the countries that produce the things we buy?
   c) for the environment?

HOW CAN WE USE MAPS TO SHOW OUR LIVES?

Bananas from the Dominican Republic

A mobile phone made in Finland

A holiday in Turkey

Trainers made in Indonesia

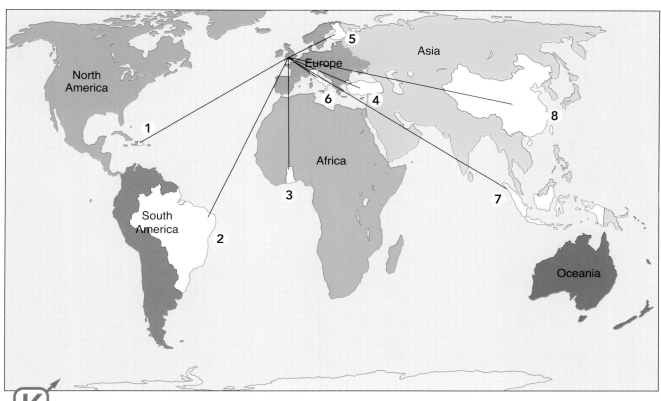

K The world, showing countries where some of the things we buy come from

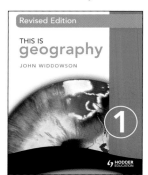
A geography book printed in Italy (you can check!)

A burger made with beef from Brazil

Chocolate made with cocoa from Ghana

A t-shirt made in China

31

your final task

You are going to produce a wall display to show your life and how it is connected to the spaces around you. The good news is that you have already done all the hard work!

What you have to do now is to bring all your work together, to make an interesting display. You could display your maps in layered circles like this. Or you could have a separate panel for each zone. You might think of another way. You may need:

- large sheets of colour paper as backing for the display
- maps at different scales
- pins or Blu-tack to stick your maps on the wall
- pencils or thread to show the connections.

Group 3 – My holiday
- You will mark places where you do holiday activities. You can use the work you did on page 27.
- Mark your places on a 1:50,000 map or draw a table to give the grid references.
- Stick your marked map in the third circle.
- Use pencil lines or thread to show the connections.

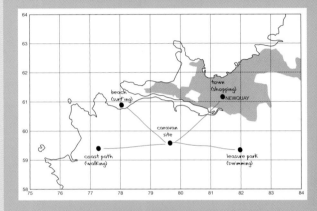

Group 1 – My school
- You will use an aerial photo or a map to show your school. You can use the work you did for page 23.
- You could put the photo or map in the middle of the display.

Group 2 – My area
- You will mark local places that you walk to. You can use the work you did on page 25.
- Mark the places on a street map and show your route using pencil lines or thread.
- Stick your marked map somewhere in the second circle.

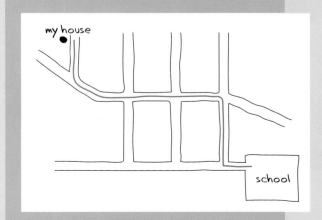

Group 4 – My country
- You will mark places where you have visited in the UK. You can use the work you did on page 28.
- Mark your places on a map of the UK.
- Stick your marked map in the fourth circle.
- Use pencil lines or thread to show the connections.

Group 5 – My world
- You will mark places where you have connections around the world. You can use the work you did on pages 30–31.
- Mark your places on a map of the world.
- Stick your marked map in the outer circle.
- Use pencil lines or thread to show the connections.

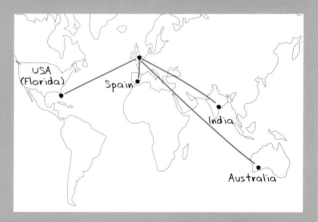

look again...

When the whole display is finished, stand back and take a good look at it. It should remind you of the work you have done in this unit.
Remember the question we started with: *How can we use maps to show our lives?* Look carefully at each part of the display. Write five sentences to say what each map tells us about your life.

3 Survivor!
Could geography help you to survive an island adventure?

KEY CONCEPTS
- **Physical processes**
- Space
- Environmental interaction and sustainable development

coming up...
Imagine you have been shipwrecked on a deserted island and you have to survive until you are rescued. You don't know how long you could be here for. How would you get on? In this unit you will find out how understanding some PHYSICAL GEOGRAPHY could help you.

through the unit...
You will keep a diary about how you survive on the island.

your final task...
Choose ten items to help you survive – and decide what will be the most helpful for you.

COULD GEOGRAPHY HELP YOU TO SURVIVE AN ISLAND ADVENTURE?

starter...

1 Your boat hits a rock and starts to sink. You have two minutes to grab some belongings before it goes down. What should you take with you? Choose just ten things from this list.

BOAT INVENTORY

compass
box of matches
pen and paper
portable TV
spade
sleeping bag
tool box (with tools)
binoculars
bars of chocolate
paint and brush
plate and cup
toilet paper
walking boots
bucket
geography book
tent
umbrella
fishing net
portable cooker
radio
map
knife and fork
sun cream
camera
toothbrush
gas bottle
axe
make-up
swimsuit/trunks
change of clothes
waterproof coat
deodorant
trainers

One thing missing from this list is a mobile phone. It would be no use anyway. Mobile phones don't work on this island!

discuss...

2 Compare the list of things you chose with a partner's list.

   a) Explain to your partner why you chose these things. For example:

   I chose the bucket because I could use it to collect water.

   b) After you have listened to each other's ideas, agree a list of the ten most important things.

Keep this list until the end of the unit. At the end you can see if you chose the most useful items.

your diary...

It's not every day you are shipwrecked. There should be plenty to put in your diary! You can use a bit of imagination. You could start like this:

It was a lovely sunny morning when I set sail from the harbour. But soon the weather began to change...

You could also include your list of rescued items in the diary.

→ Explore your island

This is your island. Of course as you drag your belongings up the beach from your wrecked boat you don't see this view! In fact, you have no idea how big this island is. Maybe, you think to yourself, there is a big town just over the hill. Or at least a lonely farmhouse with a telephone. Maybe it is not an island at all.

You set off to explore.

activity...

Work with a partner.

1 Find each of the following on the island.

> ★ a stream ★ a beach
> ★ a bay ★ a cliff
> ★ moorland ★ a lake
> ★ a forest ★ a valley
> ★ a rocky headland

2 Where do you think you would find the following features that are also on the island?

> ★ a spring ★ a rockpool
> ★ a waterfall ★ marshland

3 Where might you find these things and what do they tell you about the island?

> ★ animal droppings
> ★ a ruined cottage ★ rubbish

discuss...

4 With a partner discuss:
- Can you see any signs of life?
- Can you see anything that could spell danger?
- How are you feeling about this place?
- Will it be easy or hard to survive? Give your reasons.

your diary...

Describe your feelings as you explore your island. You could start like this:

> I'm scared. Nobody knows I'm on this island. At first I hoped that I would find some other people here. Then I discovered ...

SURVIVOR!

→ Survival challenge – read your map

A stroke of luck! You have a map that shows the island – it is called Rig Rha. The map should help you work out where you are. But there is one problem. Part of the key on the map is missing, so you need to work out what the symbols mean.

activity...

1. Study map **A**. Compare it with the drawing of the island on pages 36–37.
   a) Match the features in the key with the symbols you can see on the island.
   b) Copy and complete a key for the map to say what the symbols mean.
2. The brown lines on the map are contour lines.
   a) Compare the map closely with the drawing on pages 36–37. What do you think the lines show?
   b) Why do you only find contour lines on the map – not on the island?

(If this is difficult, look at page 39.)

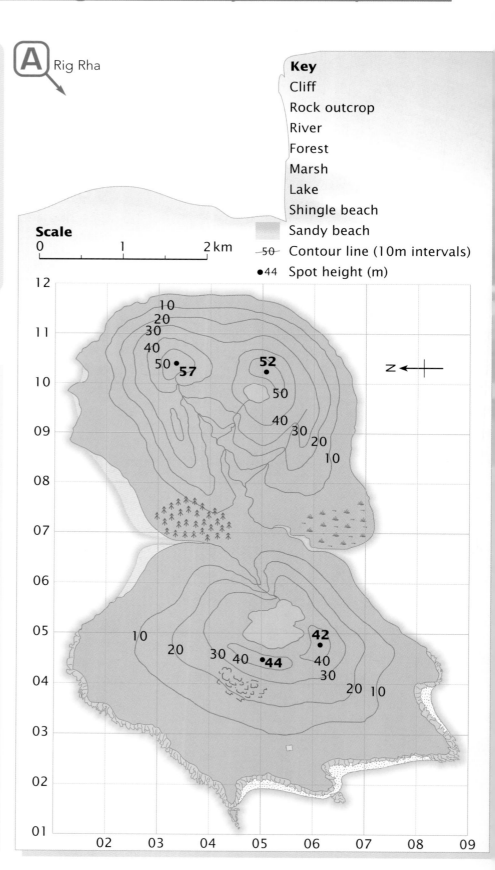

A Rig Rha

Key
Cliff
Rock outcrop
River
Forest
Marsh
Lake
Shingle beach
Sandy beach
—50— Contour line (10m intervals)
•44 Spot height (m)

COULD GEOGRAPHY HELP YOU TO SURVIVE AN ISLAND ADVENTURE?

survival challenge...

3 If a ship passes nearby you will want to attract its attention. So you need to choose a high point where you could light a fire. Find the highest point on the map.
   a) Where is it? Give a six-figure grid reference. (Look back to page 27 if you need help with grid references.)
   b) How high is it?

your diary...

4 When you get to the top of the hill you can see most of the island. Use the map to imagine what it looks like. If this is difficult, the drawing on pages 36–37 will help. Describe what you can see in your diary.

how to... Interpret contour lines

On a map CONTOUR LINES join places that have the same height. For example, anywhere on the line marked '20' is 20 metres above sea level. Of course, in real life you can't see contour lines on the ground! But if you could this is what your island would look like.

The pattern of the contours shows the shape of the land, or the RELIEF.
The main thing to know is: *the steeper the slope, the closer the contour lines.*

Your island

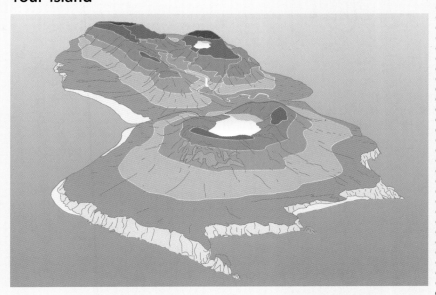

aim high...

5 Match up each of these contour patterns with one of these geographical features:
   a) gentle slope
   b) steep slope
   c) valley
   d) hill.
   Draw and label them in your book.
6 Find one example of each on the map of the island. Give a four-figure grid reference for each example.

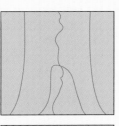
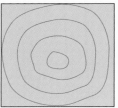

SURVIVOR!

→ Survival challenge – find fresh water

You can't live without water – so you need to find some to drink. On this island that should be easy. There is plenty of water. But, perhaps it's not so easy as you think.

Your challenge is to work out which is the **cleanest, safest water** to drink. You could just try drinking from each source and see if you get sick. However, a much quicker (and safer) way is to use your geographical knowledge. Understanding the WATER CYCLE is the key.

97% of the world's water is in the oceans, but it's too salty to drink. The other 3% is found everywhere else – in rivers and lakes, above ground and below ground, in ice at the North and South Poles, and even in the air around us.

Water is always on the move. It goes round in a never-ending cycle. As it moves from the sea to the land, through the ground and back to the sea it gets purified, then polluted, then purified, etc. Understanding the water cycle helps you know when and where the water will be at its purest.

CONDENSATION occurs when the water vapour cools and turns into water droplets.

Clouds are made up of tiny water droplets of pure water.

PRECIPITATION occurs when water droplets get too heavy, they fall as rain, hail, sleet or snow. When rain hits the ground it either flows over the ground, or sinks through the underlying rock.

WATER VAPOUR can't be seen, but it is all around you and it is pure. When water evaporates the impurities get left behind. Water vapour is not salty.

EVAPORATION occurs when the sun's heat turns sea water into vapour – it happens more quickly when it is hot.

Rivers carry the water back to the sea and pick up mud, stones and rubbish on the way.

RUN OFF is water flowing over the ground. It picks up impurities as it flows over the soil.

The **sea** may look clean, but it is salt water.

COULD GEOGRAPHY HELP YOU TO SURVIVE AN ISLAND ADVENTURE?

activity...

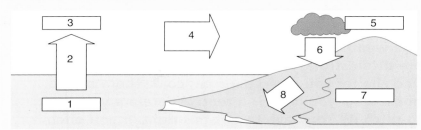

1 Look at the drawing above. It is a simplified version of the water cycle. Match each of the words below with one of the numbers on the drawing:

★ condensation ★ evaporation ★ precipitation
★ run off ★ sea ★ water vapour
★ cloud ★ river

B The water cycle on Rig Rha

survival challenge...

2 Now that you know about the water cycle, decide which of these is the best source of drinking water on this island. Make your own copy of this table to help you organise your ideas. Once you have filled out the middle three columns, use the final one to rank the sources. Give the best source a rank of 1, the worst one a rank of 5.

Source	Fresh or salty?	How reliable?	How clean?	Rank order
Sea	Salty	Will never dry up	Might be polluted	
Stream/river				
Lake				
Rainfall				
Spring				

Lakes collect water and the mud settles to the bottom. Plants, water animals and insects live in the lake.

SPRINGS form when groundwater flows out of the ground, they can keep flowing even when there is no rain.

GROUNDWATER is the water that sinks into the ground and through the rock. It is filtered clean as it slowly flows through the ground.

your diary...

Write an entry about your search for drinking water. How did you decide what the best source of water would be? Explain why you didn't choose the other sources. You could start like this:

Everywhere on the island there is water! But it's not all fit to drink.
I chose ... because ...
I did not choose ... because ...

SURVIVOR!

→ Survival challenge – find shelter

The weather on this island can be really fierce. There are often strong winds and it rains a lot. If you don't find shelter you could die from cold. There are no houses. All you have is a tent. Where are you going to pitch it? You have to get this right. If you pitch it in the wrong place it could be blown away in a gale and then you would be in a real mess.

Luckily, geography can help you with this as well. You need to use your knowledge of WEATHER and CLIMATE. Remember: weather is what happens from day to day. This can vary a lot. Climate is the average weather over many years.

C Things that affect weather and climate on the island

Sun's position The Sun is in the south at midday. In summer it rises higher in the sky and gives more heat.

Aspect This is the direction the land faces. South-facing slopes are warmer because they get more sun. North-facing slopes get less sun and are therefore colder.

Height Temperature falls as you get higher. It is colder and windier on the hilltops. Hilltops also get more rain.

Wind direction On Rig Rha wind usually blows from the west. This is the PREVAILING WIND in the UK. As it travels over the sea it picks up water vapour. So a westerly wind often brings rain.

COULD GEOGRAPHY HELP YOU TO SURVIVE AN ISLAND ADVENTURE?

Shelter East of the hills, the land is sheltered from the westerly wind. The forest also provides some shelter, but there is not much sun in the forest.

survival challenge...

1 Work with a partner.
  Look at drawing **C**. Six possible sites are marked for your tent. You have to analyse each site using the information about weather and climate on drawing **C**. Once again, you can use a table like this to help you.
  NB You might not like any of the sites. There are plenty of other possible sites on the island. You can choose a different site altogether. That's OK as long as you can prove it is a good one.

Site	Sun	Wind	Temperature	Other
1	South facing – quite sunny	Exposed to the wind	Not very warm – it's quite high	Quite near the spring
2				
3				
4				
5				
6				

your diary...

2 Write a paragraph in your diary to explain which site you chose and why. Give reasons for not choosing the other sites.

aim high...

3 The British Isles (where we live) are islands too. The same things that affect the weather and climate on Rig Rha affect our weather and climate too, but on a big scale. Try to explain these facts about the weather in the British Isles:
  a) the west side is wetter than the east
  b) the south coast gets a lot of sunshine
  c) it doesn't snow much, even in winter.

Distance from water The island is surrounded by sea. The sea warms up and cools down more slowly than the land. In summer this means the island stays cooler than you'd expect on the mainland, and in winter it means it is warmer than the mainland.

SURVIVOR!

→ Survival challenge – find food

By now you are getting hungry. You've eaten the last bar of chocolate in your pocket! You need food and you need it as soon as possible. Can geography help you with this as well? You won't be surprised that it can!

It's all about ECOSYSTEMS. Everything on the island is linked together. Plants and animals (the living things) are linked to the water, the rock, the soil, the sunlight (the non-living things) as in diagram **D**.

Ecosystems come in all shapes and sizes. There are little ecosystems all over the island: the marsh, the rockpools, the forest. The sea is a very big ecosystem. If you know where the island ecosystems are you will know where to find some food – like this seagull.

D How a rockpool ecosystem works: **a)** what's in it; **b)** what feeds on what

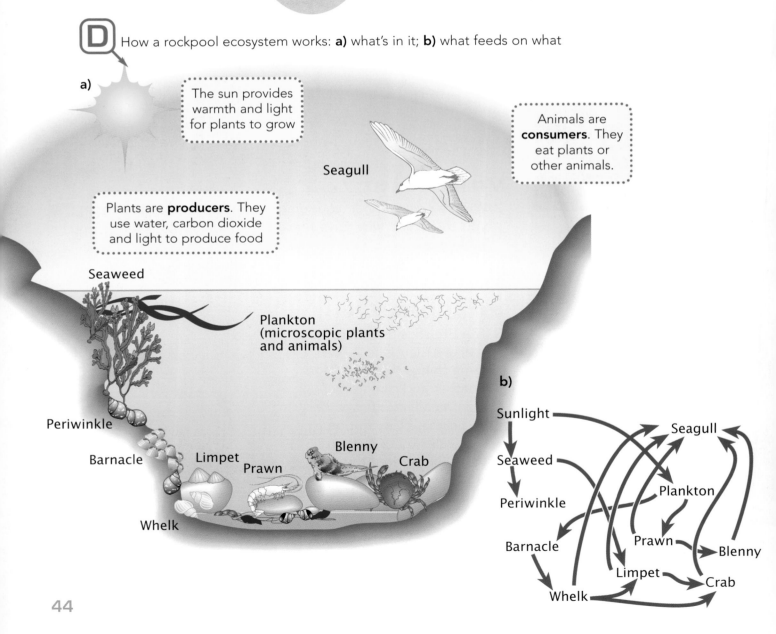

survival challenge...

1 a) Make up a menu from the plants and animals you find on the island. You need to look in the right places to find them. Here are some of the foods on the island that you could eat.

b) The ecosystems around the island include marshland, rockpools, forest, moorland and the sea. Match each ingredient with one of the ecosystems, e.g. seaweed from a rockpool.

your diary...

2 Write a paragraph in your diary to describe what food you ate, where you found it and how you cooked it. You could start like this:

Today I ate my first proper meal on the island. It was delicious! For starters I made some seaweed soup. I collected the seaweed from a rockpool. Then I...

activity...

3 Look at diagram **D**. You are now going to draw your own ecosystem, like the one shown in **Db**.

a) Copy the words below into your book. These are all parts of a moorland ecosystem. Arrange your words in a similar way.

> sunlight
> eagle
> rabbit
> insect → bird
> grass   berries
> ↙
> soil   worms

b) Underline the words in three different colours to show: producers, consumers, and the non-living environment.

c) Draw lines to show the links between each part of the ecosystem. For example: 'grass grows in the soil, birds eat insects'. Two lines have been drawn for you. Make sure your arrows go in the right direction.
When you finish you will have lots of lines. Every part of the ecosystem should be linked. Don't worry if it's a bit messy. That's what ecosystems are like!

aim high...

4 The ecosystems on this island have been undisturbed for years. Then you come along.

a) How might you change the ecosystems? (Clue: think about what you eat.)

b) If you had to stay on the island for a long time, how could you make sure you don't run out of food?

c) Could this teach you anything about the way that people should live on Earth?

SURVIVOR!

→ Rescued!

Luckily, you didn't have to spend years on the island after all! A passing boat saw your smoke and reported it to the local coastguard. A helicopter was sent to pick you up. You survived your island adventure.

your final task...

When you landed on the island you didn't know what to expect. You had just two minutes to grab things from the boat before it sank. There was no time to think properly.

Now, you know more about the island, if you could choose again, what would you take? Work with a partner.

1 Look at this inventory again.
   a) Choose ten items that you would take. This time think very carefully. Write a list.
   b) Give a reason for taking each item.
2 Compare your list with the items that you chose at the start of the unit. Did your list change? If so,
   a) how, and b) why did it change?
3 In real life do you think that geography would help you survive?
   a) What were the most important things that you learned to help you survive?
   b) What else would you need to know to help you?

BOAT INVENTORY

- compass
- box of matches
- pen and paper
- portable TV
- spade
- sleeping bag
- tool box (with tools)
- binoculars
- bars of chocolate
- paint and brush
- plate and cup
- toilet paper
- walking boots
- bucket
- geography book
- tent
- umbrella
- fishing net
- portable cooker
- radio
- map
- knife and fork
- sun cream
- camera
- toothbrush
- gas bottle
- axe
- make-up
- swimsuit/trunks
- change of clothes
- waterproof coat
- deodorant
- trainers

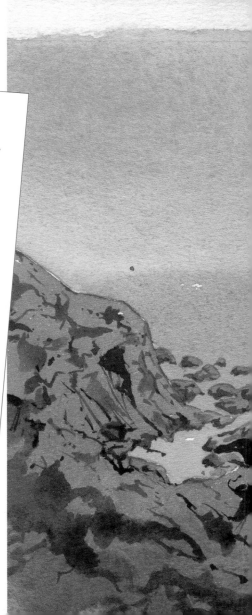

SURVIVOR!

→ Saved by geography!

The island of Rig Rha is made up, although it is based on a real island in the Outer Hebrides. As far as we know, no one has ever been shipwrecked there in real life – but in future, if they are, perhaps they should learn some geography first!

But here is a *true* story. It was published in The *Sun* newspaper soon after the tsunami (giant wave) that hit south-east Asia in December, 2004.

GEOGRAPHY LESSON SAVES TILLY, 10, AND FAMILY

An alert 10-year-old British girl saved her family and 100 other tourists from the devastating Asian tsunami – because she had learnt about the giant waves in a geography lesson.

She explained she had just completed a school project on the huge waves and they were seeing the warning signs that a tsunami was minutes away. Her parents alerted the other holidaymakers and staff at their hotel, which was quickly evacuated.

The wave crashed a few minutes later, but no one on the beach was killed or seriously injured.

Tilly Smith was holidaying with her mother Penny, father Colin and seven-year-old sister Holly on Maikhao beach in Phuket, Thailand, when the tide suddenly rushed out.

As the other tourists watched in amazement, the water began to bubble and boats on the horizon started to violently bob up and down.

Tilly, who had studied tsunamis in a geography class just two weeks earlier, quickly realised they were in terrible danger. She told her mother they had to get off the beach immediately and warned there could be a tsunami.

aim high...

Write your own newspaper account about how geography helped you to survive your island adventure. It might not be quite so dramatic as Tilly's story, but you can still make it interesting. Try to include all the different ways in which geography helped you.

You will study more physical geography through your course. You will learn about earthquakes and tsunamis in Book 3 of *This is Geography*.

Tilly, from Surrey, gave the credit to her geography teacher, Andrew Kearney, at Oxshott's Danes Hill Prep School. She told *The Sun*, 'Last term Mr Kearney taught us about earthquakes and how they can cause tsunamis. I was on the beach when the water started to go funny. There were bubbles and the tide went out all of a sudden. I recognised what was happening and had a feeling that there was going to be a tsunami. I told mummy.'

Tilly's mother said she was 'very proud' of her daughter, while her headteacher Robin Parfitt said she had 'wisdom beyond her years'.

Mr Kearney said he remembered teaching Tilly and her fellow students that after the sea was sucked backwards, the next five to ten minutes were crucial for people to survive. He said her quick-witted actions were typical. 'I'm stunned at the news but so relieved she and her family are safe,' he told *The Sun*. 'Tilly is a very bright level-headed girl. Nothing illustrates her character more than her brave actions in a terrifying situation.'

SURVIVOR!

→ What have you learned so far?

Through the first three units you have been learning to use various geographical skills. Complete a copy of the table below to show how you used these skills.

Skills I have learned to use
Use symbols and a key
Draw symbols on a plan
Draw a simple scale plan
Interpret an aerial photo
Measure distance on a map
Follow/describe a route
Use grid references
Use an atlas to find places
Interpret contour lines

COULD GEOGRAPHY HELP YOU TO SURVIVE AN ISLAND ADVENTURE?

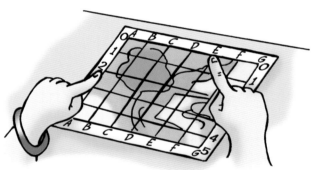

How I have used them
I measured distance from my primary school

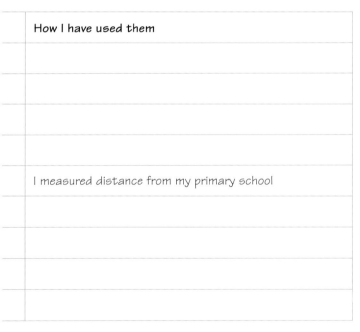

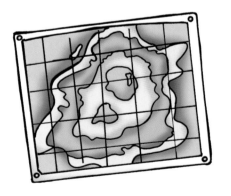

4 City – past, present, future

Why do people choose to live in cities?

KEY CONCEPT

- Human processes
- Place
- Diversity

A

B

In the UK, most people live in cities. Over 1 million people live in this one. What do you think of it? Would you like to live here?

C Manchester

D

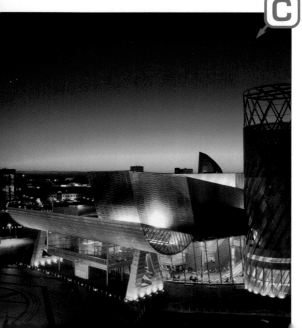

starter...

1 Look carefully at the eight photos of Manchester. Make two lists –
 • things you like about the city
 • things you dislike about the city.

activity...

2 Why do people come to Manchester? One way to find out is to look in their bags!
  a) Examine the contents of this bag. Who do you think it belongs to?
  b) Explain your choice. Why have they got these things in their bag?

a shopper
a business person
a tourist
a runaway teenager

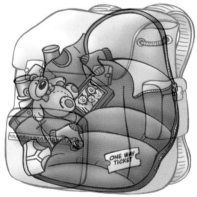

aim high...

3 Why do you think this person came to Manchester? Make up a story about their first day in the city. Try to include places shown in the photos, in your story.

coming up...

You are going to visit one city – Manchester. You will find out why people came to live here years ago, and what still brings them here today. You will find out how the city has changed, and is still changing now.

your final task...

You will use your own ideas *and* what you have found out about Manchester to suggest how to improve a problem area of a city – so that people will still want to come and live there in the future.

→ The Romans arrive

Manchester started as a Roman SETTLEMENT. A settlement is a place where people live.

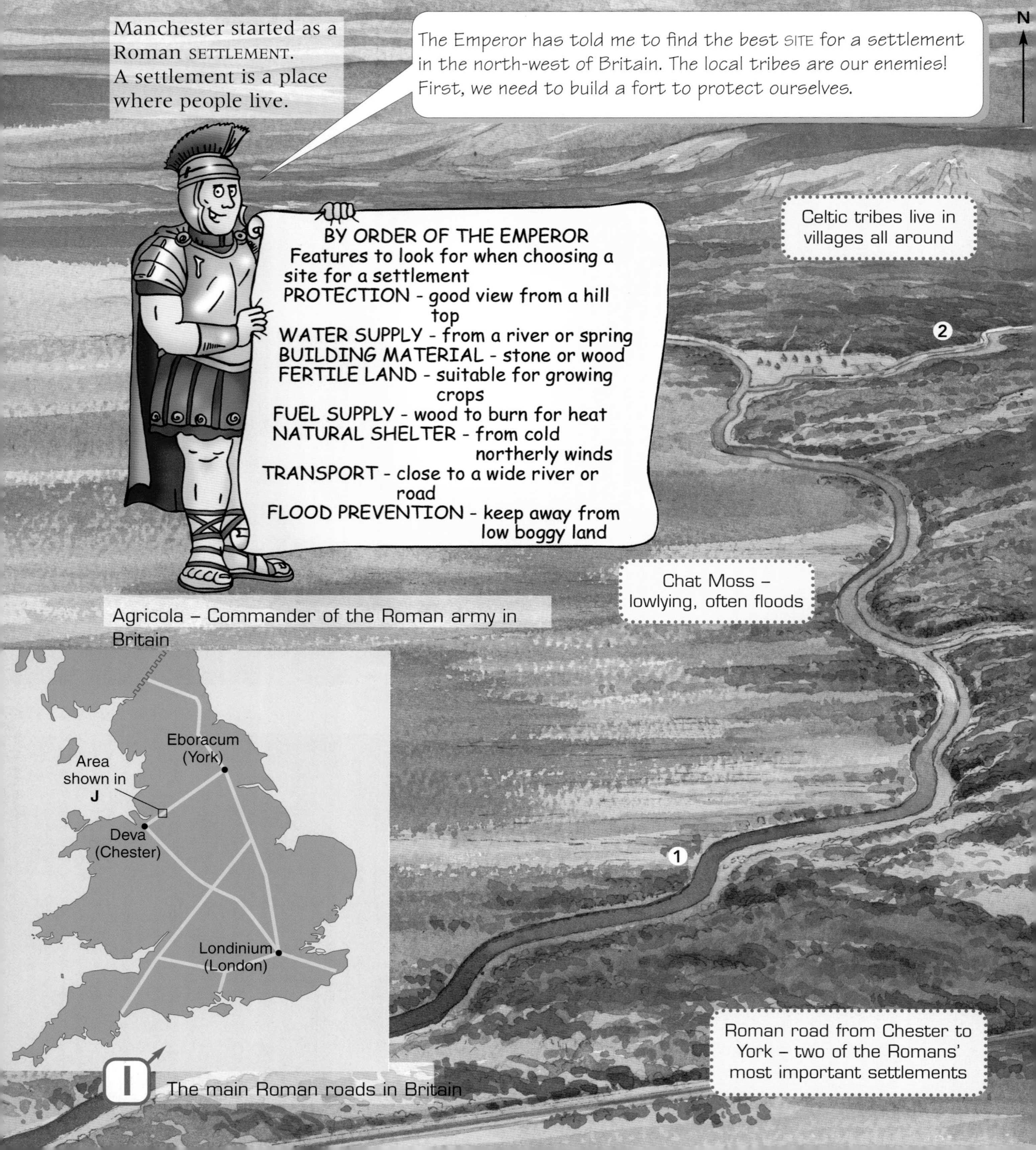

The Emperor has told me to find the best SITE for a settlement in the north-west of Britain. The local tribes are our enemies! First, we need to build a fort to protect ourselves.

BY ORDER OF THE EMPEROR
Features to look for when choosing a site for a settlement
PROTECTION - good view from a hill top
WATER SUPPLY - from a river or spring
BUILDING MATERIAL - stone or wood
FERTILE LAND - suitable for growing crops
FUEL SUPPLY - wood to burn for heat
NATURAL SHELTER - from cold northerly winds
TRANSPORT - close to a wide river or road
FLOOD PREVENTION - keep away from low boggy land

Agricola – Commander of the Roman army in Britain

Celtic tribes live in villages all around

Chat Moss – lowlying, often floods

Roman road from Chester to York – two of the Romans' most important settlements

Area shown in J

Eboracum (York)
Deva (Chester)
Londinium (London)

I The main Roman roads in Britain

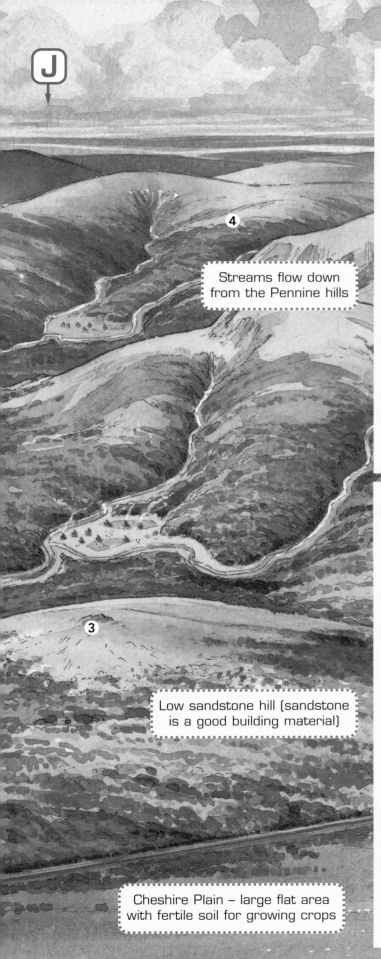

J

Streams flow down from the Pennine hills

Low sandstone hill (sandstone is a good building material)

Cheshire Plain – large flat area with fertile soil for growing crops

activity...

1 You have to choose the best site for a Roman settlement. There are four possible sites, numbered 1 to 4 on drawing **J**. Work with a partner if possible.

Stage 1 Draw a large table, like the one below, to help you to make a decision.

Features needed for a settlement	Site 1	Site 2	Site 3	Site 4
Protection				

In the first column list the features needed for a settlement.

Stage 2 Fill in your table. Look at drawing **J** carefully. Tick the box if a site has that feature nearby.

Stage 3 Choose the best site. The best site should be the one with most ticks.

aim high...

Stage 4 Draw a sketch map to send to the Emperor to show at least three good things about the site that you chose. Label these important features on your map. You can start your sketch map like this:

The Romans called their settlement Mancunio. The English word for a Roman fort is 'chester'. That is where the name, Manchester, comes from. Did you choose the same site as Agricola? Your teacher can tell you which site Agricola chose.

CITY – PAST, PRESENT, FUTURE

→ Manchester grows

After the Romans left not a lot happened in Manchester for the next thousand years! Through the Middle Ages it was a small MARKET TOWN where people came to buy and sell. It was after 1750, during the INDUSTRIAL REVOLUTION, that Manchester really took off.

1750

Manchester had **cotton** mills. Fast-flowing rivers provided water power to run the machines in the mills.

1765

Water mills were replaced by steam power. A **canal** was built to bring coal to Manchester to fuel the steam engines. This was the first canal in Britain.

1804

Canals were a great success. Manchester was now at the heart of Britain's canal network. Goods and raw materials were **traded** around the country.

activity...

1. Draw a timeline from 1750 to 1900. Mark six important dates on the line and label what Manchester was like at the time. Here are some words you could use:

   canal   cotton mills   industrial city   port   railway   trade

   For example: Manchester was a small town with cotton mills

   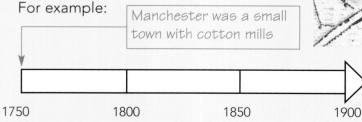
   1750   1800   1850   1900

K Manchester in 1650

2. Look at map **K**.
   a) On a copy of the map, label these features: a river; a bridge; flat farm land; roads to other towns.
   b) Explain how each of the features that you have labelled might have helped Manchester to grow.

WHY DO PEOPLE CHOOSE TO LIVE IN CITIES?

aim high...

3 Compare map **K** and map **L**. Map **L** shows part of the same area 200 years later. (Be careful! Map **L** has a larger scale. Although it is bigger it covers a smaller area than map **K**.)

Describe at least five differences between the two maps. Write a sentence about each one. For example:

In 1650 there was farmland in the area, but by 1850 there was none.

1830
Railways are the future!

The first **railway** from Manchester to Liverpool opened. Now, more goods could be moved, more quickly, and people could travel further to work.

1850
We can hardly breathe down here!

There were factories and mills all over Manchester. It was now a large **industrial city** – and very polluted.

1894
Next stop America

The Manchester Ship Canal opened, linking Manchester with the Irish Sea and the rest of the world. Manchester became a **port**. Goods could now be traded directly with other countries.

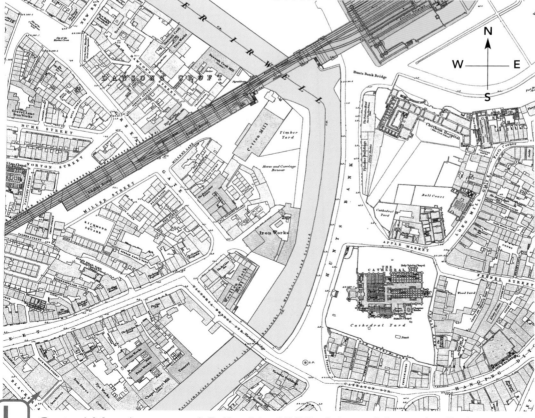

Key
- River Irwell
- railway
- factories and mills

L Central Manchester and Salford in 1850. Adapted from the Ordnance Survey map

CITY – PAST, PRESENT, FUTURE

➜ More work, more people

By 1850, Manchester was one of the largest cities in England. Factories attracted people from nearby towns and villages to work there. Some people came from much further away.

As industry grew, so did Manchester's population (see table **N**). The same was happening all over England. Diagram **M** shows how the proportion of people living in cities grew during the nineteenth century. This process is called URBANISATION.

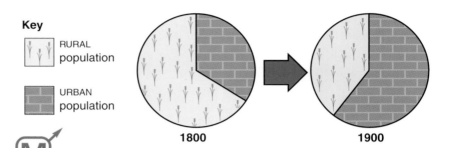

M Urbanisation in England in the nineteenth century

Year	Population
1701	around 10,000
1751	17,000
1801	75,000
1851	303,000
1901	544,000
1951	703,000
2001	393,000

 Population change in Manchester

You might wonder if cities still grow like this today. They don't in this country. Most of our cities have stopped growing and some are losing people. But some cities in Africa and Asia are growing as fast as cities in England did 100 years ago.

Coming from the Mill – a painting by L.S. Lowry in 1930. Lowry lived in Manchester. He painted hundreds of scenes in the city in the early twentieth century. This was when the cotton industry, and Manchester's population, reached their peak.

WHY DO PEOPLE CHOOSE TO LIVE IN CITIES?

activity...

1 Make a copy of diagram **M**. Now, use the diagram to explain what urbanisation is. Write one or two sentences, including these words:

process   urban   proportion

2 Draw a line graph to show the change in Manchester's population.
   a) Draw two axes to fill a page in your book. First, turn your book sideways as your graph will need to be wider than the one here. Number each axis as shown below. Make sure the numbers are equally spaced.

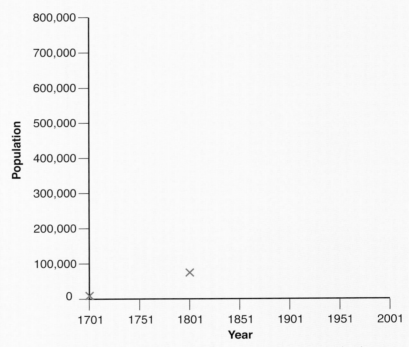

   b) Look at table **N**. Mark the population at each date in the table on your graph in pencil with a cross. Join the crosses with a line.
   Keep this graph – you will use it again on page 62.

3 Describe the line on the graph you have drawn. Write three sentences, beginning:

At first, after 1701...
After 1801, the population began to...
Since 1951, it has...

4 Look at painting **O** with a partner.
Think of words to describe the scene in Lowry's painting. Choose both positive and negative words, e.g. busy (+ve), polluted (–ve). **Make two lists.**

CITY – PAST, PRESENT, FUTURE

→ Multicultural Manchester

After 1950, mills started to close and the population declined. Since then Manchester's population has been falling. Most of the old industries, like cotton, have gone. Workers have moved out of the city to find other jobs.

In their place new groups of people have arrived. Many are IMMIGRANTS from other countries. They came here for work and a better quality of life. Today, Manchester is a MULTICULTURAL city made up of different communities that have settled there.

P Communities in Manchester

Irish people first came in the mid-nineteenth century to escape famine and poverty in Ireland. Today, one-third of the city's population is of Irish descent.

Chinese people have come since the 1970s. Chinatown is a thriving part of central Manchester, with its own shops and businesses.

West Indian people came in the 1950s. They were needed to work in services, like health and transport.

Asian people, mainly from India and Pakistan, first came to work in the cotton and textile industries in the 1950s. As factories closed they found other jobs.

Jewish people came in the nineteenth century to escape persecution in Europe. Manchester still has a Jewish community today.

Jaydev was born in Manchester to Asian parents. He tells his story about growing up in the city. This is a true story. It was first told in the Moving Stories Exhibition at the Peoples' History Museum in Manchester.

JAYDEV'S STORY

My dad was the very first Asian to arrive in Stalybridge (on the edge of Manchester). He came in 1956. The British wanted people to come over to work. My dad was taken in by the propaganda. You know – the streets are paved with gold, and all that. What he found when he got here was totally different really. He was an engineer back home. He came and ended up working in a cotton mill. Then he became a bus driver. He couldn't get work that he was trained to do.

He came here with the view that he was going to make a better life for his family. Not just for his wife and kids here, but also his family back home, because he had to send a lot of money back home.

With a lot of our parents, when they came here they felt threatened. So they went deeper into their religion and culture. They were very strict with us kids. I remember once, I used to hang out on the streets a lot. I was just talking to this girl. And I was smoking. We were teenagers. My dad was driving his bus. He saw me. He had lots of passengers. He just stopped the bus, got off, dragged me on and drove me home. That has got to be the most embarrassing moment of my life. My dad was ranting on the bus as he was driving and it was full of white people.

Now I would describe myself as British Asian. There's a lot of things about the British way of life I like. And there's a lot of things about the Asian way of life I like. You try to live both. I can slip into either quite easily.

Today, Rusholme (near the city centre) is the focal point for the Asians in Manchester. People come here from all over the place, especially at occasions like Eid. They call it the curry mile. I think the wider community just views us as curry and bhangra music. But we're a lot deeper than that.

activity...

1 Answer these questions about immigration.
   - Why come to Britain?
   - Why move to a city, like Manchester?
   - Why live together in areas, like Rusholme?
   - What benefits do immigrants bring to Manchester?

   Read Jaydev's story to help you to answer the questions.

aim high...

2 Read Jaydev's story again, carefully.
   a) Work out the questions that the interviewer may have asked.
   b) What other questions would you like to ask Jaydev?
   c) Prepare questions for an interview with someone who has experienced moving (they don't have to be an immigrant). Carry out your interview and record the answers.

CITY – PAST, PRESENT, FUTURE

→ Manchester living graph

On this spread you need to think back to all you have found out about Manchester so far. Some of the information to help you is on this page. The rest is dotted through pages 52–61.

Q Copy of Manchester population graph

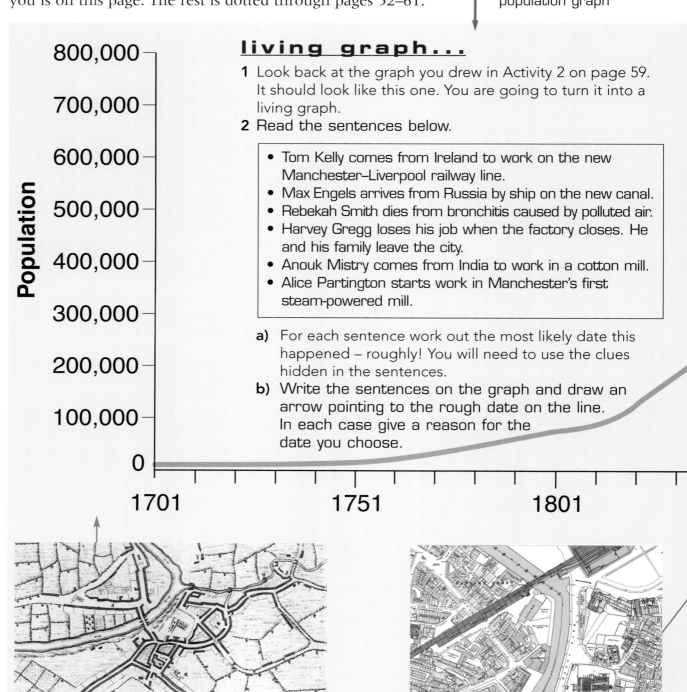

living graph...

1 Look back at the graph you drew in Activity 2 on page 59. It should look like this one. You are going to turn it into a living graph.
2 Read the sentences below.

- Tom Kelly comes from Ireland to work on the new Manchester–Liverpool railway line.
- Max Engels arrives from Russia by ship on the new canal.
- Rebekah Smith dies from bronchitis caused by polluted air.
- Harvey Gregg loses his job when the factory closes. He and his family leave the city.
- Anouk Mistry comes from India to work in a cotton mill.
- Alice Partington starts work in Manchester's first steam-powered mill.

a) For each sentence work out the most likely date this happened – roughly! You will need to use the clues hidden in the sentences.
b) Write the sentences on the graph and draw an arrow pointing to the rough date on the line. In each case give a reason for the date you choose.

aim high...

3 Now try these phrases. They are harder. Be sure you can think of a reason.

- The hardest time to live in Manchester.
- The best time to find a job in Manchester.
- The easiest time to buy a house in Manchester.
- The most exciting time to come to Manchester.

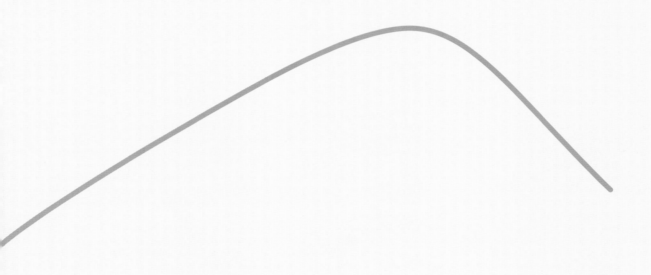

1851　　　　　1901　　　　　1951　　　　　2001

Year

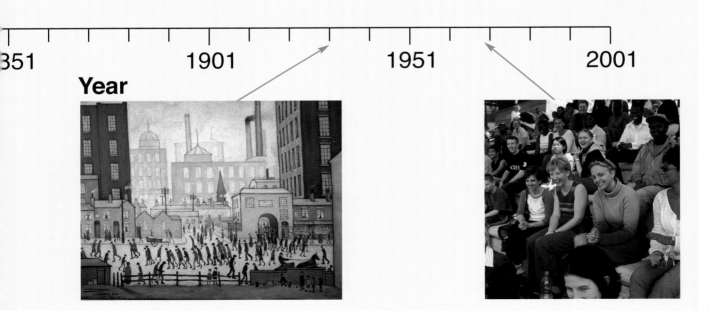

CITY – PAST, PRESENT, FUTURE

→ Manchester re-invented

Today, Manchester no longer has many of the FUNCTIONS that it had in the past.

- The city does not need a fort to defend it any more.
- People no longer come to buy and sell cows and pigs at the market.
- The port has closed.
- Most of the old industry has gone.

So what functions does Manchester have today? Quite a few, actually!

R Manchester city centre

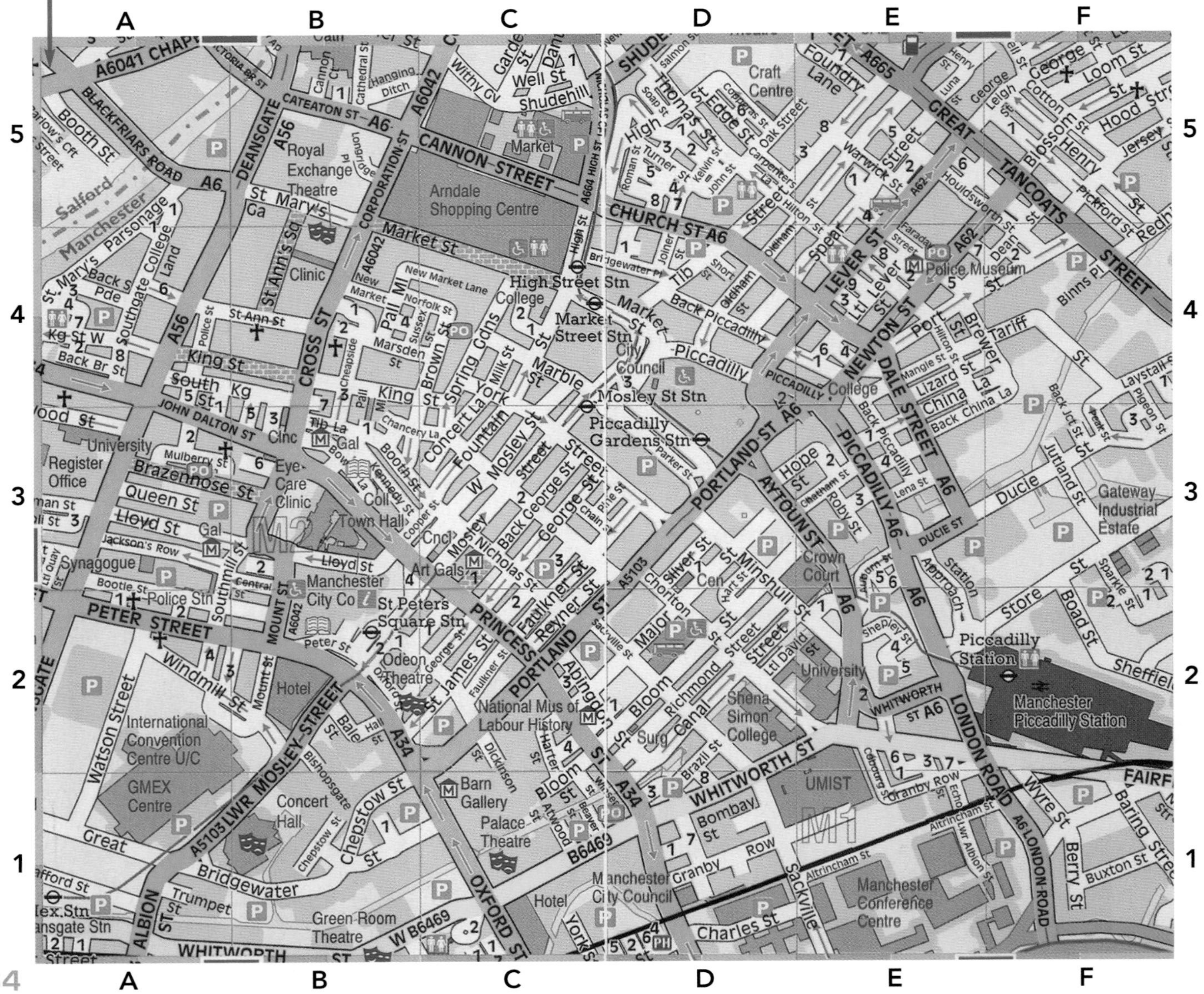

WHY DO PEOPLE CHOOSE TO LIVE IN CITIES?

activity...

1 a) Look for evidence on map **R** for each function listed in the box below.

> ★ tourism ★ government
> ★ education ★ entertainment
> ★ shopping ★ business
> ★ industry ★ transport centre

b) Complete a table like the one below. For each function give the map evidence and the map square in which you found it.
An example is done for you.

Function	Evidence	Map reference
Government	city council	D1/D4

2 The drawings in **S** show the changing fortunes of a warehouse building in Manchester. (A warehouse stores goods or raw materials for industry.)
  a) What uses did the warehouse have in 1800, 1950 and 2000?
  b) Write some sentences to explain the changes from what you already know about Manchester.

aim high...

Tourism is one of Manchester's functions today.

3 Plan a weekend trip to Manchester, using map **R**.
  a) How would you get to the city?
  b) Where would you stay?
  c) What would you visit? (If you don't fancy any of the places on the map, then turn over to pages 66–67 for more ideas.)
  d) How would you get around?
  Write details of your plans.

S The changing fortunes of a warehouse in Manchester

➜ Twenty-first century Manchester

Manchester is moving on. Over the past twenty years there has been massive REGENERATION in and around the city centre. Much of the city has been rebuilt as old industries closed and old buildings were demolished. The aim is to create a city that people will want to live in, where businesses will want to invest and people will want to visit. Many other cities in the UK are being regenerated in the same way.

Salford Quays is an area around the old docks on the Manchester Ship Canal. It became derelict when the docks closed in 1980. Since then, new housing, offices and shops have been built. Two new attractions – the Lowry Arts Centre and the Imperial War Museum – now bring thousands of visitors to the area. Salford Quays is only ten minutes from the city centre by MetroLink.

Trafford Park is still an industrial area. But the old factories have been knocked down and replaced by modern buildings that are cheap for companies to rent. There are plenty of jobs. Nearby is Old Trafford, Manchester United's stadium. The football club also employs many people.

activity...

1 Work in a group of three.
   a) Each person in the group should choose one of these roles:
      - a person thinking about where to live
      - a tourist deciding which city to visit
      - a company director thinking about where to locate their business.

      What would make your person want to come to Manchester?

      Read all the information on these two pages. Look for improvements in Manchester that would persuade your person to come to the city. For example, the company director might be attracted by the modern offices. Try to find at least three reasons for each person. Make a list.
   b) Compare your list with the rest of your group. Agree a list of the five most important improvements. (These will be the improvements identified by most people in the group.)

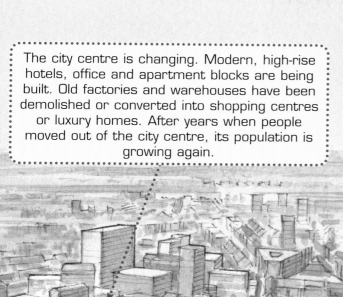

The city centre is changing. Modern, high-rise hotels, office and apartment blocks are being built. Old factories and warehouses have been demolished or converted into shopping centres or luxury homes. After years when people moved out of the city centre, its population is growing again.

aim high...

2 Design a leaflet to attract *either* new residents and new business *or* tourists, to Manchester. It should mention all the main improvements brought about by regeneration in the city. Think about the people that your brochure is aimed at. Mention the improvements that you think will particularly attract this group.

Sportcity was built for the 2002 Commonwealth Games. It has many sports and leisure venues. It includes the City of Manchester stadium, now home to Manchester City football club. New homes, shops and offices have been built overlooking the old canals. East Manchester used to be one of the poorest areas of Manchester. Now it's nicknamed 'Venice of the North'.

MetroLink is a fast, convenient tram system that uses both roads and disused railway track. It links the city centre with places around Manchester, making it easier for people to travel without their cars.

Hulme, south of the city centre, was an area of high-rise council flats built in the 1960s. People did not want to live there because of the crime and vandalism. Now, the high-rise blocks have been demolished and replaced by houses with gardens. Crime levels have fallen.

CITY – PAST, PRESENT, FUTURE

→ Improve Metropolis

■ your final task...

The story of Manchester is like the story of many cities in the UK. The same changes that happened there have happened in other cities too.
Here is a city that needs a few of the improvements that Manchester has had. See if you can plan a better future for Metropolis.
Can you make it a city where people want to live?

METROPOLIS PRESENT – SPOT THE PROBLEMS

The numbers on drawing **T** highlight some of the problems in the area today.
Match the numbers with each of the comments below.

Metropolis

METROPOLIS FUTURE – PLAN A BETTER CITY

Think of ways to tackle each problem. You could base your ideas on what you saw in Manchester (pages 66–67) or think of your own ideas.
Draw your plans for the area. Your teacher will give you a large copy of this drawing to do it on.

a) Many of the buildings, the canal, railway and roads are still there. Can you think of better ways to use them? Draw your ideas.

b) How could you make better use of the spaces that are left? Draw your ideas. (Don't worry if your drawing is not perfect. It's the ideas that are important.)

c) Annotate your plans on the drawing. Describe and explain the improvements that you have made. For example:

> I have knocked down the high rise flats because no one likes living there. There's too much crime and vandalism.

"It's such a shame to see that lovely old building lying empty and unused."

"It takes me so long to drive anywhere. And then I can never find a place to park."

"There's not enough space for us to play around here."

5 Shop until you drop!

How is the way that we shop changing?

KEY CONCEPT
- **Space**
- Interdependence

coming up...

In this unit you are going to meet some of the characters who live in the fictional village of Ogden Bridge in the Lake District. There have been big changes in Ogden Bridge recently. Grocer Jack used to run the village store, but as you can see, now he's working as a lollipop man. In this unit you will find out what has happened to the shops in Ogden Bridge, why Grocer Jack has had to change jobs, and how similar things have been happening in villages throughout the UK.

you could...

Similar changes may have happened where you live. You could do a shopping investigation in your local area to find out how we shop in real life.

your final task...

At the end of the unit you will write your own up-to-date children's story about shopping.

starter...

Work with a partner to try to solve a mystery! Why has Grocer Jack become a lollipop man?

a) Read all the clues below. Your teacher may give you them on cards.
b) Try to find connections between the clues.
c) Place clues that are connected to each other close together. Be careful – you may not need to use all the clues.
d) Stick the clues on the page. Draw lines between the clues that are connected.
e) Make up a story to explain what has happened. Include some of the characters introduced in the clues. You will meet some of them again later.

Ogden Bridge is a small village with 400 people.

Gemma Gates and Kevin Kettle are best friends. They go to secondary school in Kendal.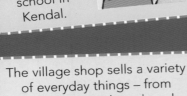

The Kettles are a busy family. They do one big shop every week to save time.

A new superstore has opened near Kendal, a town about 5km away.

Grocer Jack has been running the village shop in Ogden Bridge for years.

Julie is the local vicar. She is amazed by the new superstore. 'The prices are so low and there is so much choice.'

The village shop sells a variety of everyday things – from newspapers to bread, and baked beans to soap.

Some houses in Ogden Bridge now lie empty for most of the year.

The new superstore has a large car park and is easy to drive to.

Betty Button has left the village. She sold her house to a family from London to use as a holiday home.

The number of people who live in Ogden Bridge is falling.

Grocer Jack decides that he cannot afford to run the village shop any more.

 Fred Jessup used to buy a daily paper. Now he watches the news on TV.

There was an armed robbery at the store last Christmas. Jack's arm was broken!

Shops in Ogden Bridge and nearby villages have closed one by one over the past few years.

Ogden Bridge Rovers lose 16–1 to Kendal United.

Ms Meek used to ride a scooter. Now she has learnt to drive a car.

At rush hour the road outside the village school is very busy. One child has been injured this year.

SHOP UNTIL YOU DROP!

How we used to shop

Grocer Jack has had to find a new job as a result of the changes in the way we all shop these days. Although Ogden Bridge is a fictional place, similar things have been happening to *real* people in *real* places. Read article **A** and interview **B** to find out more.

A — From Reuters News Agency, 19 June 2003

'POSTMAN PAT' POST OFFICE SHUTS ITS DOORS

The rural English post office that provided the inspiration for children's character Postman Pat closed its doors for the last time yesterday after a slump in business. The post office in the village of Beast Banks, near Kendal in the Lake District, was made famous by author John Cunliffe in his stories about a cheery postman.

Beast Banks became the home of Postman Pat in 1981 after resident author, John Cunliffe, was asked to write a series of short stories for the BBC. He said that the post office's closure was a sad day for the village. 'The post office is a very important part of the community. You can exchange gossip, buy things and get personal help there. The spirit of the village will go.'

B — An interview with Grocer Jack

Grocer Jack, why has the village shop closed?

It just wasn't making money any more. There weren't enough customers. Lots of people go to Kendal now to do their shopping. And there aren't so many people in the village as there used to be.

How do you feel about it closing?

It's a terrible shame. The village shop has been my life. Closing down is the hardest decision I've ever made. What will Ogden Bridge be like without a shop? There'll be nowhere left for people to go for a chat. It was the last shop left in the village. Now they've all gone. And it's the same in Crosthwaite, Grigghall and all the little villages around here.

What shops did the village used to have?

When I was a young lad there was a butcher's, a baker's and a little shop where you could buy fancy gifts. People hardly ever went to Kendal in those days.

How did shopping then compare to now?

We never complained. There wasn't much choice, but at least everything was on our doorstep. I know people say that it's easier to shop in a supermarket, but it's a twenty-minute ride on the bus to Kendal, and the bus only comes twice a day. It's alright if you have a car, but some people don't.

Where do people shop these days?

Some go to the high street in Kendal. They still have ordinary shops. If they are like me, they want to shop where the people know you and you get service with a smile. I don't like supermarkets. Everybody always seems to be in such a rush.

C A typical British high street grocer's shop in the early twentieth century, outside ... and inside

activity...

1 Look at the photos in **C**.
   a) Compare the traditional grocer's shop in **C** with a modern supermarket that you know. Using evidence in the photos, write five sentences to compare:

   - **size** A modern supermarket is ... than a traditional grocer's shop.
   - **location** A modern supermarket is located ... whereas ...
   - **type and range of goods** A modern supermarket sells ... while ...
   - **number of customers** More customers use ... compared with ...
   - **how people get there** Most people travel to a modern supermarket by ... However ...

   b) Where would you prefer to shop? Explain why.
2 Interview an older person that you know, to ask about how shopping has changed. It could be a parent, a grandparent or a teacher.
   - **Before the interview** Prepare some questions for the interview. You could use questions like the interview with Grocer Jack, or make up your own.
   - **During the interview** Listen carefully to the answers. Make notes about what they tell you.

aim high...

3 **After the interview** *Either* write up the interview neatly with questions and answers, *or* write a short magazine article about how shopping has changed.

SHOP UNTIL YOU DROP!

→ Let's go shopping

Where do you like to shop? Round the corner, in town, out of town, on the internet? Where you choose will depend on what you want to buy.

People go shopping for two main types of goods:

- CONVENIENCE GOODS are things that we need every day like milk, bread or newspapers. They need to be convenient to buy.
- COMPARISON GOODS are things that we buy less often but cost more money, like clothes, electronic goods or furniture. We like to compare prices and styles before we buy them.

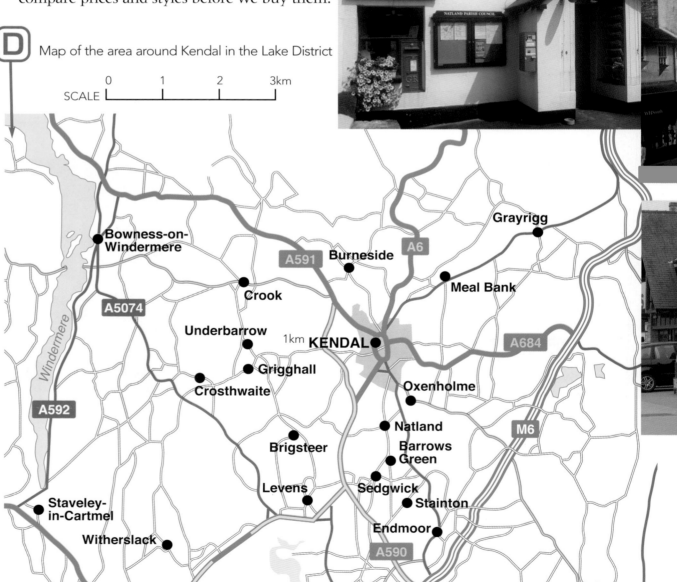

D Map of the area around Kendal in the Lake District

E Small village store in Natland. Ideal for things that you need to buy every day

HOW IS THE WAY THAT WE SHOP CHANGING?

activity...

1. Fred Jessup has written a shopping list.
   a) Classify the items on the list into convenience goods and comparison goods.
   b) Suggest where Fred used to go to buy each item on the list (before the village shop closed) – a village store, a high street shopping centre or an out-of-town superstore.

> pint of milk
> new pair of glasses
> cat food
> book next year's holiday
> apples
> wife's birthday present
> mobile phone
> newspaper
> new shoes
> cream cake

2. The opening of the new out-of-town superstore at Kendal is one reason that the village shops in the district have closed. Suggest what effect the superstore will have on:
   a) shoppers who drive a car
   b) shoppers without a car
   c) high street shops in Kendal.

3. Read these statements about shopping centres. Do you think they are true or false?

   A People make more frequent trips to small shopping centres than large shopping centres.

   B People travel further to small shopping centres than to large shopping centres.

   C People are more likely to buy comparison goods in small shopping centres than large shopping centres.

   D People are more likely to go by car to large shopping centres than small shopping centres.

   a) Copy the statements you think are true.
   b) Change the statements you think are false. Rewrite them as true statements.

   Keep the statements that you have written. These are your hypotheses. An HYPOTHESIS is an idea that you want to test.

F High street shopping centre in Kendal. A range of shops where you can go to do your regular shopping

G Large out-of-town shopping centre outside Kendal. You'll find everything on one site. But it helps if you've got a car to get there. This has a large Morrison's supermarket as well as the stores shown in the photo

coming up...

Over the next four pages you are going to carry out a shopping investigation to see if your hypotheses are true or false.

SHOP UNTIL YOU DROP!

→ Investigate how people shop

LOCAL INVESTIGATION

investigate...

Look at the hypotheses that you wrote about shopping centres in Activity 3 on page 75. Now you are going to carry out an investigation to test your hypotheses.

First, you will do a shopping survey at a *small* local shopping centre (like your local high street) to find out how people shop there. Later, you will compare it with our results from a *large*, out-of-town, shopping centre.

Step 1 Collect information

At the shopping centre
Work with a partner. You will need a clipboard, pencil and a blank copy of shopping questionnaire **I**.

Do a shopping survey using the questionnaire in **I**. Interview ten shoppers. Take turns with your partner to ask the questions and to record the answers. Use the ten columns to record answers from the ten shoppers. Tick the correct answer box for each question.

Step 2 Enter your data into a spreadsheet

Share your results with the rest of the class. To do this, enter your survey results onto a computer spreadsheet like the one shown in **J**. The whole class should enter results on the same spreadsheet.

H

Doing a shopping survey. Always remember to:
- Work in pairs
- Be polite
- Keep it short
- Record the answers

HOW IS THE WAY THAT WE SHOP CHANGING?

I A shopping questionnaire carried out in a large shopping centre. You will use a blank version of the same questionnaire so that, later, you can compare results. You will complete all the blue bits yourself.

SHOPPING QUESTIONNAIRE										
Location: Kendal Superstore Date/Time: 1 May 2008										
	1	2	3	4	5	6	7	8	9	10
A How often do you shop here?										
a) daily				✓						
b) weekly, or more often		✓	✓				✓		✓	✓
c) monthly, or more often	✓				✓			✓		
d) less often						✓				
B How far did you travel to get here?										
a) less than 1 km				✓						
b) 1 to 5 km				✓			✓			✓
c) 5 to 10 km	✓	✓						✓	✓	
d) over 10 km					✓	✓				
C What transport did you use to get here?										
a) car	✓	✓		✓	✓	✓			✓	
b) bus							✓			✓
c) taxi								✓		
d) walk or cycle			✓							
D What are you shopping for?										
a) food				✓			✓			✓
b) clothes or shoes			✓					✓		
c) household/electrical goods					✓					
d) gifts									✓	
e) more than one of the above	✓	✓				✓				

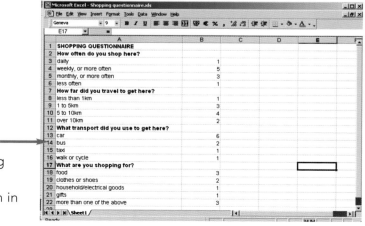

J Results from a shopping survey at a large shopping centre, shown in an Excel spreadsheet

SHOP UNTIL YOU DROP!

→ Compare shopping centres

LOCAL INVESTIGATION

The table and the pie charts below show the results from our shopping survey at a *large* shopping centre.

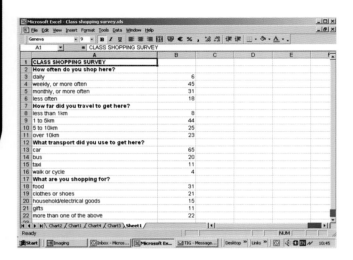

What are you shopping for?

- food
- clothes or shoes
- household/electrical goods
- gifts
- more than one of the above

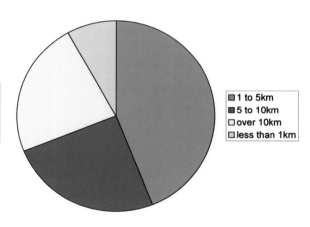

How far did you travel to get here?

- 1 to 5km
- 5 to 10km
- over 10km
- less than 1km

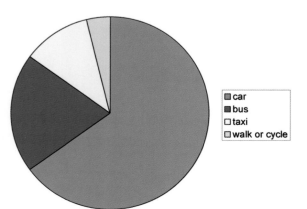

What transport did you use to get here?

- car
- bus
- taxi
- walk or cycle

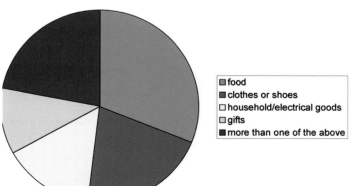

What are you shopping for?

- food
- clothes or shoes
- household/electrical goods
- gifts
- more than one of the above

78

investigate...

Step 3 Present your results

Create four pie charts from your spreadsheet, like the ones we created on page 78. Each pie chart shows the answers for a question in the shopping survey. Your teacher may give you a sheet that explains how to create pie charts from a spreadsheet. Alternatively, you could draw your own pie charts by following the instructions below.

Draw a pie chart

1. Draw a circle. Divide the circle into quarters with a pencil.

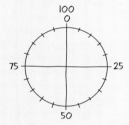

2. Write percentages around the pie chart: 0%, 25%, 50%, 75% and 100%. Mark 5% intervals around the pie chart.

3. Divide the pie chart into segments to show your results. Use the % marks around the edge as a guide to make it accurate.
   - If your results add up to 100 it is easy to draw the segments – the figures in your results are already percentages.
   - If your results don't add up to 100, you need to turn them into percentages. For example, if you interview 50 shoppers and 9 of them shop daily, this is how you work out the percentage:

   $\frac{9}{50} \times 100 = 18\%$.

   Use the percentages to draw the segments.

Step 4 Write your conclusions

Here are the hypotheses that you read on page 75. Did you rewrite any of them?

A. People make more frequent trips to small shopping centres than large shopping centres.

B. People travel further to small shopping centres than to large shopping centres.

C. People are more likely to buy comparison goods in small shopping centres than large shopping centres.

D. People are more likely to go by car to large shopping centres than small shopping centres.

Now, you are going to test the hypotheses using the results of your investigation. You visited a *small* shopping centre. You will compare it with a *large* shopping centre. To test the hypotheses:

a) Match the pie charts that you created in Step 3 with the pie charts on page 78. For example, match the charts for the question: 'How often do you shop here?'.

b) Look at the charts to prove whether each hypothesis is true or false. For example, if the first hypothesis is true then more people will go daily to the small shopping centre than to the large shopping centre.

c) Write a conclusion to your investigation. Did you prove whether each hypothesis is true or false? For example:

> Our investigation proved that hypothesis A is true. More people travel daily to small shopping centres than to large shopping centres.

At the end of your conclusion, rewrite any hypothesis that is false.

SHOP UNTIL YOU DROP!

→ Internet shopping

Now there's another way to shop – on the internet. Julie the vicar in Ogden Bridge has discovered that she can order goods on-line and have them delivered to her door. But, before she trusts it, first she wants to know how the internet works.

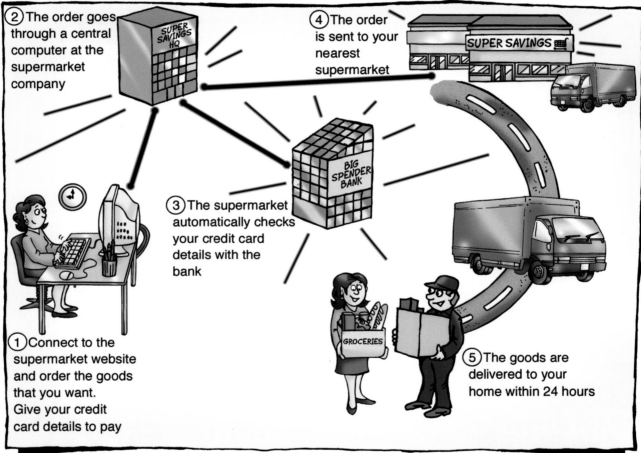

K How internet shopping works

activity a...

1 Work with a partner.
   a) Look carefully at diagram **K** (your partner should not look). Explain to your partner how internet shopping works.

   b) Now, without looking at diagram **K**, can your partner explain the process of internet shopping back to you? How could you have described it better?

   c) Why are some people anxious about shopping on the internet?

activity b...

2 a) Write a shopping list with at least six items. Include at least three convenience goods and three comparison goods. Could you do all your shopping on the internet?

b) Find out where you could buy each item on the list by checking internet websites. A good website to start with is www.e-shopping.co.uk/shopping.htm. It will direct you to other shopping websites.

c) Many people use the internet now to do their research before they buy. How does the internet help them do this?

d) Is internet shopping more useful for convenience or comparison shopping?

discuss...

3 Talk with a partner.
a) Do either of you have any experience of shopping on-line? What did you buy? How successful was it? Were there any problems?

b) How do you think internet shopping compares with ordinary shopping? Think about which offers more choice, more convenience, costs less and has least impact on the environment.

4 Draw a table like the one below. List at least three advantages and three disadvantages of shopping on-line.

Advantages	Disadvantages
It is more convenient to stay at home	

L A Tesco van delivering shopping ordered on-line

SHOP UNTIL YOU DROP!

your final task...

Our fictional account of the goings on in Ogden Bridge was inspired by the real-life closure of the Postman Pat post office in the Lake District. As you have found out, there have been a lot of changes to the way we shop over the past few years:

- many small shops (and post offices) have closed
- there are more out-of-town shopping centres and superstores
- many people drive cars
- many people have computers at home and use the internet
- it is possible to shop on-line and have shopping delivered.

Using all of the information you have found out about in this unit you are now going to write a new up-to-date children's story about shopping.

1 Look at each of the scenes 1–4 below. For each one:
   a) Write down where the characters are shopping. You can choose from: village, high street, superstore or on-line shopping at home.
   b) Work out what problem is being experienced and why.
   c) List possible solutions to the problem.

See, it didn't look like that when I ordered it!

I'm sure there used to be a sports shop here!

2 Now choose just one of the situations shown and write your own children's story around that situation. Don't forget to:
  a) introduce your characters and a setting – it will be more effective if you set it somewhere you know so that you can add interesting details
  b) describe the problem your main character faces
  c) add an ending to your story to solve the problem.

3 Try to include some of these shopping words in your story:

  comparison   convenience
  out-of-town shopping   high street
  internet   on-line shopping
  home delivery   village store

4 Finally, give your story a title.

6 Flood disaster
How could we be better prepared next time?

KEY CONCEPTS
Environmental interaction and sustainable development
Space
Physical processes

Summer, 2007, was the wettest ever recorded in England. Rainfall was more than twice the summer average. Large areas of the country were flooded after torrential rain.

FLOODS seem to be happening more often in Britain these days. Scientists put this down to CLIMATE CHANGE as a result of global warming (find out more in *This is Geography Book 2*). If they are right, we need to be prepared for more floods in the future.

Tewkesbury Abbey, Gloucestershire, in July 2007

HOW COULD WE BE BETTER PREPARED NEXT TIME?

Monday 23 July 2007

Flood Misery in Gloucestershire

Britain's worst summer floods in living memory have forced hundreds of families in Gloucestershire to leave their homes and have caused chaos throughout the region. Roads have been closed, railway services suspended and, despite being surrounded by water, thousands of people have no drinking water.

More than a month's worth of rain fell in a few hours last Friday. Since then, river levels in central England have been rising, flooding parts of Worcestershire and Gloucestershire up to depths of two metres.

Tewkesbury, one of the worst hit towns, where the River Severn and River Avon meet, has been virtually cut off by floodwater. Water poured into the town, flooding over five hundred homes and shops. Dozens of families were evacuated. Last night across Gloucestershire, more than two thousand people were sleeping in temporary shelters.

Emergency services were busy through the weekend, rescuing people from their homes by boat. RAF helicopters were called into action to lift people stranded in outlying farms.

A water treatment works on the River Severn, near Tewkesbury, has been flooded, cutting off water supplies to 150,000 people in the region. It is too early to say what the eventual cost of the floods will be, but insurance companies are predicting that claims will run into billions of pounds.

coming up...

Have you ever experienced a flood? Hopefully, you never will. It is hard to imagine the misery and devastation that a flood brings.

In this unit you will come as close as possible to the experience of a flood – without actually getting wet!

through the unit...

You will create a concept map about flooding showing the causes, the effects and the responses.

your final task...

At the end of the unit you will design a poster to warn people about the risk of flooding, so that next time we will be better prepared.

starter...

1 Read the newspaper article carefully. Identify:
  a) the causes of the flood   b) the effects of the flood
  c) the responses to the flood.
  Underline a copy of the article using three colours to highlight the causes, effects and responses.

2 You are going to create a concept map to show the causes, effects and responses for the flood disaster.
  Use a full page in your book. In the centre of the page write the title 'FLOOD DISASTER'. Around it write the three headings: 'Causes', 'Effects', 'Responses', like this:

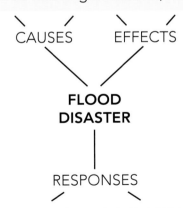

Now write the ideas that you found in the article around the three headings. Link them to the correct heading. As you go through the rest of the unit add more causes, effects and responses to your concept map. You may be able to group your ideas into sub-headings. For example, causes can be grouped into physical and human causes. By the end your page should be full.

FLOOD DISASTER

→ A flood diary

Simon Walden lives with his family in Tewkesbury. He kept a diary during the floods. On Friday 20th July he was driving home after the rain. Simon and his family were lucky that their home didn't get flooded. Many other families were not so lucky.

A Tewkesbury town centre during the flood

Friday, 8.00p.m. I made it to Evesham (fifteen miles from Tewkesbury), which is completely surrounded by floodwater. They're stopping all traffic about a mile down the main road. Decided to try my luck on the smaller roads.

10.00p.m. Finally made it to the outskirts of Tewkesbury, but could not do the final 300 yards to get home. I abandoned the car and crossed the bridge over the river on foot. As long as I gripped a fence, wall or lamppost and – crucially – did not lift my feet off the ground, I was OK.

2.00a.m. A helicopter is hovering over our house. My son, Sam, and I go out and follow the searchlights. We see the bridge wall has collapsed. Cars are floating away. The house is still four feet above water level.

Saturday, 7.00a.m. Flood waters much higher – now only a couple of feet from the house. Go to town to buy some bread and milk. Everyone says it's never been like this before. By the time we get home we feel reasonably certain that this is the worst of it.

6.00p.m. It's getting worse! A fire engine is parked outside the house. We're told that this is the last chance to EVACUATE; after that it will be boat or helicopter, and only in an emergency.

2.00a.m, 4.00a.m, 6.00a.m. Get up at two-hourly intervals to check the water level. It is still rising and we've got about six inches safety margin before the garage floods.

Sunday, 8.00a.m. Bleary-eyed and beleaguered. The sun is shining and it's almost pleasant, except for the water level, which inches closer. There are helicopters galore – some TV news, some rescue. Our house is on TV all the time.

12p.m. Sam and I go into town. It's very quiet, no cars at all. All the shops are closed. We help at the abbey, which is flooding, by carting a ton of sandbags to the great oak doors and moving all the chairs. The vicar tells us we've earned our place in heaven!

10p.m. Water level hasn't changed all day.

Monday, 8.00a.m. Water level still hasn't changed. Bottled water is delivered to the bridge, and we get some by wading and carting.

4.00p.m. We're told we could lose power and the phone line at any moment. We're out of bread, baked beans, and are low on milk.

6.00p.m. We've been to the town centre. It's very sad to see the shops and houses flooded. I saw a chap sitting on his doorstep looking very down. He'd only had a couple of inches of water in the house, but that's all it takes to completely ruin a home.

10.00p.m. TV news likes to talk it up; words like 'pandemonium' are just nonsense. People here are very calm and chatty – swapping gossip, news and encouragement and helping neighbours. There's no more tap water but we still have heat and power.

B What a mess! Flood damage inside a house

activity...

1 How would you deal with a flood? Read the list of things you could do in the box below.

> Listen to flood warnings on TV or radio
> Evacuate your house
> Open all the doors and windows
> Turn off electricity, gas and water supplies
> Call your insurance company
> Move furniture, carpets and valuable items upstairs
> Don't try to walk or drive in floodwater
> Put sandbags down outside doors

a) Put these things into order: actions you could take **before**, **during**, and **after** a flood.
b) Give a reason for carrying out each action.

2 Imagine the damage a flood could cause in your home. Photo **B** will help you. This house has been flooded to a depth of one metre. Think of all the things in your home below this depth. Make a list of the damage and problems that a flood would cause.

3 Look at the contents of this flood kit. For each item in the kit, write a sentence to explain how it might be useful. For example, *Candles would give some light if the electricity is switched off.*

remember!

Add more causes, effects and responses to your concept map.

FLOOD DISASTER

→ Background to the flood

Tewkesbury is a small town in Gloucestershire. It lies near the confluence – or meeting point – of the River Severn and the River Avon. Find the rivers on map **D**.

The oldest part of the town, around the medieval abbey, was built on a small hill rising above the FLOODPLAIN – the flat land on either side of a river. This helped to keep it dry when the rivers flooded. That was until 2007, when even the abbey got wet.

Over the years the town expanded as the population grew. Today, Tewkesbury has 10,000 people.

C Tewkesbury Abbey with the River Avon behind it

activity...

1 Find Tewkesbury Abbey on map **D**. It is at grid reference 891 324.
   a) Why do you think the town was built so close to the river? Give two reasons. (Clue: look back at page 54.)
   b) What is the problem with this site? (Clue: look back at page 84.)
2 Contour lines on the map show the shape of the land. Anywhere on flat land (with no contours), close to a river, is in danger of flooding. Find the following places on the map. In each case, say whether they are at risk of flooding.
   a) Saxon's Lode at 864 388
   b) The church in Church End at 894 361
   c) Tewkesbury Golf Club at 881 312
   d) Mythe Water Treatment Works (Wks) at 890 336

aim high...

3 During the 2007 flood, Tewkesbury was cut off. Look carefully at map **D**. For each of the following routes, explain why it flooded. Use grid references to help you. For example,
   The M50 crosses the River Severn and its flood plain at 866 368.
   a) The M5 motorway
   b) The A38 main road
   c) The railway line

HOW COULD WE BE BETTER PREPARED NEXT TIME?

D 1:50,000 Ordnance Survey map extract of Tewkesbury and the Rivers Severn and Avon

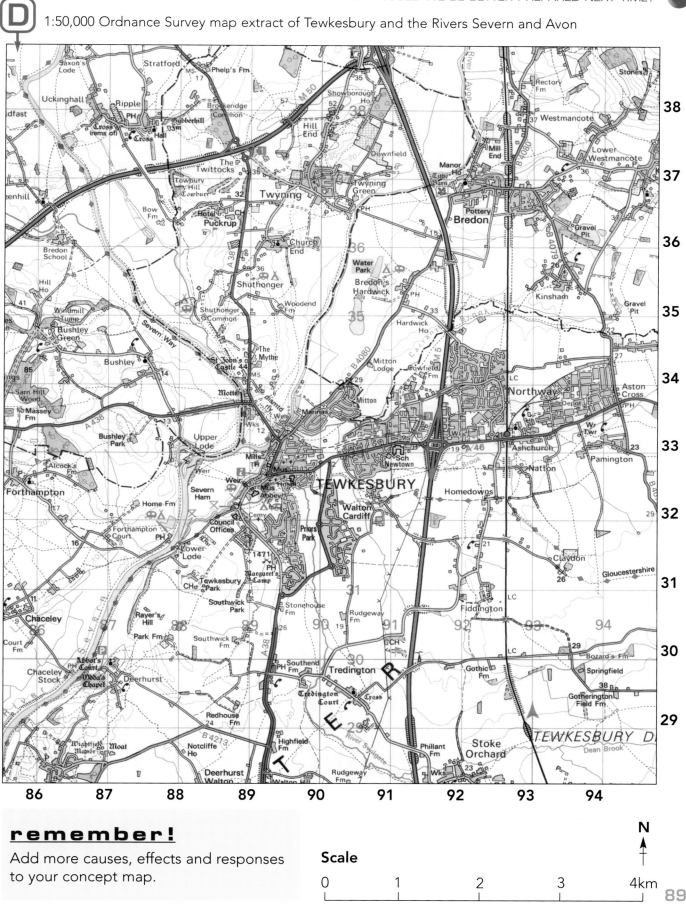

remember!
Add more causes, effects and responses to your concept map.

Scale: 0 1 2 3 4km

N

FLOOD DISASTER
➡ Rivers and flooding

River flooding happens where there is more water flowing in a river than its CHANNEL can hold. Water overflows and spreads across the floodplain. Drawing **E** shows the factors that affect the risk of flooding.

Factors that reduce the risk of flooding

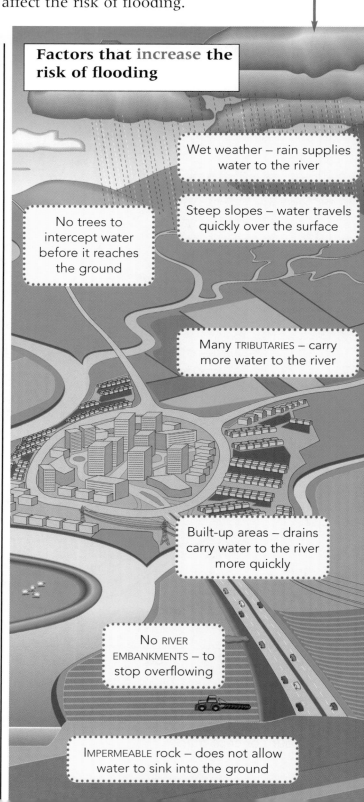

Factors that increase the risk of flooding

- Wet weather – rain supplies water to the river
- Steep slopes – water travels quickly over the surface
- No trees to intercept water before it reaches the ground
- Many TRIBUTARIES – carry more water to the river
- Built-up areas – drains carry water to the river more quickly
- No RIVER EMBANKMENTS – to stop overflowing
- IMPERMEABLE rock – does not allow water to sink into the ground

HOW COULD WE BE BETTER PREPARED NEXT TIME?

The ENVIRONMENT AGENCY looks after the main rivers in England and Wales. It monitors rainfall and river levels 24 hours a day. When floods are forecast it issues warnings on radio, TV and the internet, and advises people on what to do. Its website has flood maps for every area of the country, showing the flood risk (map **F**).

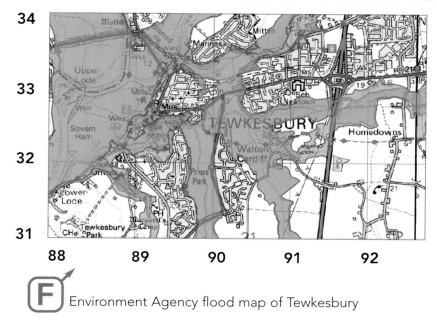

F Environment Agency flood map of Tewkesbury

Key
- Area at risk from a normal flood (say, every twenty years)
- Area at risk from an extreme flood (say, every hundred years)

activity...

1 Look at drawing **E**. Read the factors that increase the risk of flooding on the right. Notice the differences between the two sides of the drawing.
   a) Make a list of the factors on the left side of the drawing that would reduce the risk of flooding. They are the opposites of the factors on the right. For example, *dry weather*.
   b) For each factor, explain why it would reduce the risk of flooding. For example, *dry weather – little rain to supply water to the river*.

2 Look at map **F**.
   a) Find each of these places: the Abbey, Prior Park, Council Offices, the hospital.
   b) For each of these places decide:
      i) if it is in danger of flooding or not
      ii) how frequently could it flood?

3 Here are four ways in which the Environment Agency helps to reduce the risk of flooding:
   a) it monitors river levels and gives flood warnings
   b) it builds flood defences, such as embankments
   c) it plants trees near the river
   d) it recommends no more building on floodplains.
   In each case, write a sentence to explain how it helps to reduce the flood risk.

aim high...

4 Find out if your home is at risk from flooding. Go to the Environment Agency web site at www.environment-agency.gov.uk/flood. Type in your postcode to see your local flood map.

remember!

Add more causes, effects and responses to your concept map.

FLOOD DISASTER
→ Tewkesbury under water

Places look different when they are under water. Can you recognise any of the features around Tewkesbury in photo **G**?

HOW COULD WE BE BETTER PREPARED NEXT TIME?

activity...

1 Look at photo **G**.
   a) Compare the photo with map **D** on page 89 and map **F** on page 91. (It is easiest to work with a partner and share books.)
   b) Match numbers 1–6 in the photo with the following places:

   Tewkesbury town centre

   Tewkesbury Abbey

   River Severn

   River Avon

   Priors Park

   Mythe Water Treatment Works

2 a) In which direction was the camera pointing in photo **G**? (Clue: north is at the top of map **D**.)
   b) Describe the pattern of flooding around Tewkesbury. You can use these sentence starters to help you.

   The worst affected areas seem to be ...

   The east of the town ...

   Most of the town centre ...

remember!

Add more causes, effects and responses to your concept map.

G Tewkesbury and surrounding area in July 2007

FLOOD DISASTER

→ Floodwater rising

The floods in Tewkesbury followed the normal pattern. After the rain, river levels began to rise (graph **I**). They continued to rise until the rivers burst their banks and flooded the town. The delay between when the rain fell and the floods arrived gave the Environment Agency enough time to put out flood warnings. Consequently, no one in Tewkesbury drowned.

H Water rising at a road intersection in Tewkesbury

I Rainfall figures and river levels on the River Avon near Tewkesbury in July, 2007

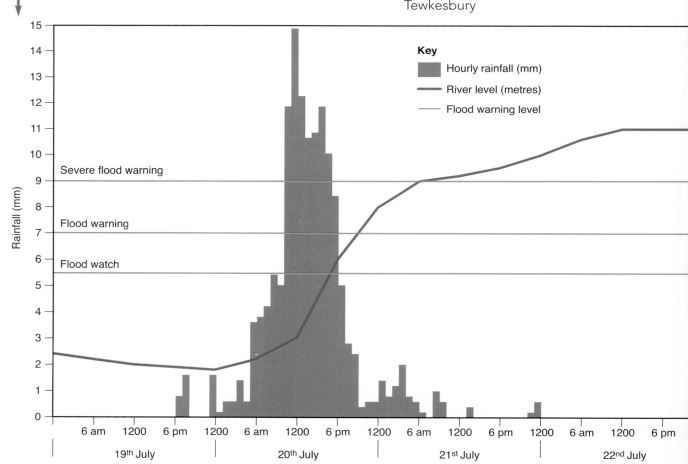

HOW COULD WE BE BETTER PREPARED NEXT TIME?

activity...

1 Study graph **I** carefully. Look at the rainfall bars.
   a) On which day did most rain fall?
   b) What was the highest hourly rainfall?
   c) At what time of day was this?
2 Look at the line showing river level in graph **I**.
   a) On which day did the river reach its highest level?
   b) What height did the river reach?
   c) How long after the highest rainfall did this happen?
3 a) What connection can you see between rainfall and the river level on the graph?
   b) How can you explain the connection?

aim high...

4 You are working for the Environment Agency in July, 2007. Use graph **I** to decide when each of these flood warnings should be given in Tewkesbury:
   a) Flood watch
   b) Flood warning
   c) Severe flood warning.
Give the day and time when each warning should be given.

remember!

Add more causes, effects and responses to your concept map.

FLOOD WARNING CODES

Flood Watch Flooding is possible. Be aware! Be prepared! Watch out!

Flood Warning Flooding of homes, businesses and main roads is expected. Act now!

Severe Flood Warning Severe flooding is expected. Imminent danger to life and property. Act now!

All Clear Flood watches and warning no longer in force. Seek advice to return.

Flood Disaster

→ The big picture

Tewkesbury was not the only place under water in July, 2007. Several towns along the Rivers Severn and Avon flooded. Others would have suffered if flood defences had not been in place (see photo **K**). The Environment Agency gave out 55 flood warnings in England and Wales, including eight severe flood warnings.

The floods caused other problems. The closure of Mythe Water Treatment Works meant that 150,000 homes in Gloucestershire were without any water supply. People had to collect their water from tanks in the street (see photo **L**). In Gloucester an electricity substation flooded, cutting off power to 50,000 homes.

activity...

1 Look at map **J**. Tewkesbury was one of the towns worst affected by the 2007 floods. Why do you think this was? (Clue: look at Tewkesbury's position on the map.)

2 Study map **J** carefully. You work for the Environment Agency. After the 2007 floods you are asked to advise the government on how some of the problems experienced in the 2007 floods could be avoided in future. For example, you could advise – *Build better flood defences in towns like Upton and Tewkesbury to protect them from flooding.* Make at least three more recommendations.

discuss...

3 Some people think that it is not worth trying to protect towns like Tewkesbury because the flood risk is so high and the cost of protection would be so great. Instead, they say, people should move away from the town. What do you think?

remember!

Add more causes, effects and responses to your concept map.

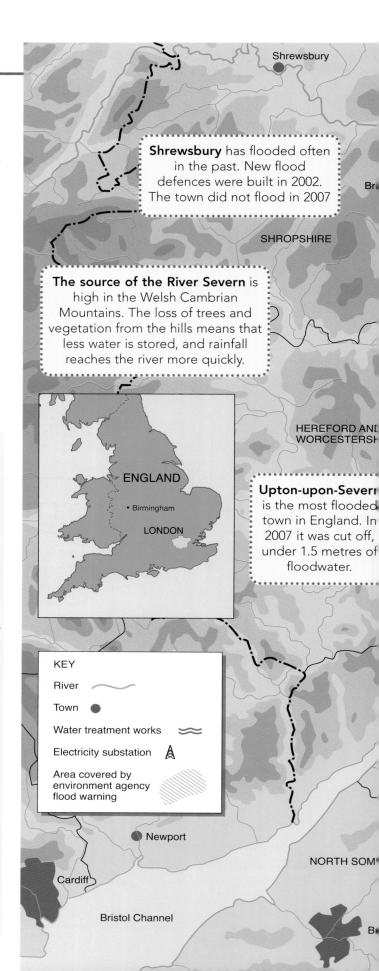

Shrewsbury has flooded often in the past. New flood defences were built in 2002. The town did not flood in 2007

The source of the River Severn is high in the Welsh Cambrian Mountains. The loss of trees and vegetation from the hills means that less water is stored, and rainfall reaches the river more quickly.

Upton-upon-Severn is the most flooded town in England. In 2007 it was cut off, under 1.5 metres of floodwater.

KEY
River
Town
Water treatment works
Electricity substation
Area covered by environment agency flood warning

HOW COULD WE BE BETTER PREPARED NEXT TIME?

K Building flood defences in Shrewsbury

L People collecting water from tanks in Gloucester

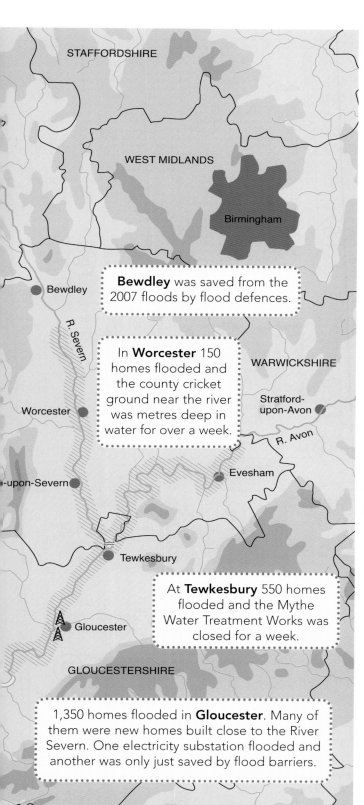

Bewdley was saved from the 2007 floods by flood defences.

In **Worcester** 150 homes flooded and the county cricket ground near the river was metres deep in water for over a week.

At **Tewkesbury** 550 homes flooded and the Mythe Water Treatment Works was closed for a week.

1,350 homes flooded in **Gloucester**. Many of them were new homes built close to the River Severn. One electricity substation flooded and another was only just saved by flood barriers.

The mouth of the River Severn is tidal as far upriver as Gloucester. High tides can add to flooding problems.

J Map of the River Severn, showing some of the causes, effects and responses to flooding

FLOOD DISASTER

activity...

Look at the concept map that you made. Compare it with the one below. There are lots of ways to organise the information. Yours may be just as good. Check that you included all the causes, effects and responses on your concept map. Use the page numbers (shown in circles) to go back and find them for yourself.

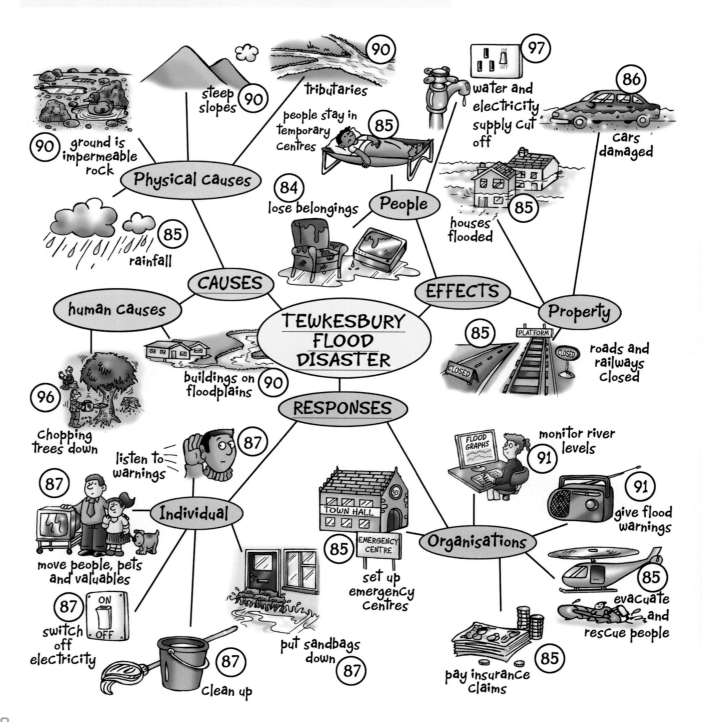

HOW COULD WE BE BETTER PREPARED NEXT TIME?

→ Be prepared for flooding

Across the UK the cost of the floods in 2007 was huge. But the cost was not just about money. Many homes and businesses were ruined, treasured possessions were lost and lives were disrupted. Eleven people were killed by floods. The question is, how can we be better prepared next time it happens?

your final task...

If we are going to be better prepared the next time a flood happens, people need more information. You have been asked by the Environment Agency to design a poster to tell people about the risk of flooding.

Use the concept map to help you to design your poster.

- First, decide whether people need to know more about the causes, effects or responses to flooding. For example, you might decide they need to know more about the effects.
- Next, choose which idea, or ideas, on the concept map is most important for people to know. Don't try to put too many ideas on your poster. For example, you could choose the idea that people can lose their belongings in a flood.
- Finally, design your poster. You can use a desk-top publishing package to help you to design it on a computer. You could choose any of the photos or drawings in this unit to download. Alternatively, you could do your own drawing, or find an image on the internet. Add your own writing to the poster to get your message across.

Here is a poster designed by the Environment Agency. Could you do even better?

7 What a load of rubbish!
What should we do with all our waste?

KEY CONCEPT
- Environmental interaction and sustainable development
- Scale
- Diversity

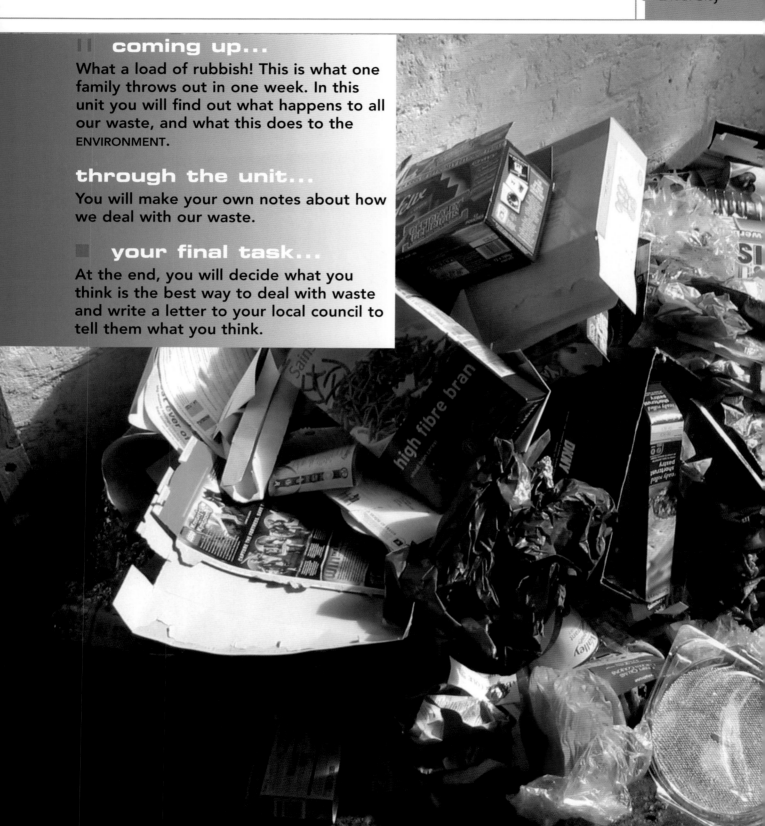

coming up...
What a load of rubbish! This is what one family throws out in one week. In this unit you will find out what happens to all our waste, and what this does to the ENVIRONMENT.

through the unit...
You will make your own notes about how we deal with our waste.

your final task...
At the end, you will decide what you think is the best way to deal with waste and write a letter to your local council to tell them what you think.

WHAT SHOULD WE DO WITH ALL OUR WASTE?

starter...

1 First, think about the different types of waste we produce. Look at the rubbish thrown out by one family, below.

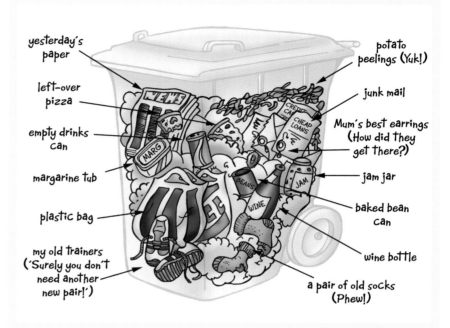

- yesterday's paper
- left-over pizza
- empty drinks can
- margarine tub
- plastic bag
- my old trainers ('Surely you don't need another new pair!')
- potato peelings (Yuk!)
- junk mail
- Mum's best earrings (How did they get there?)
- jam jar
- baked bean can
- wine bottle
- a pair of old socks (Phew!)

a) Think of five or six headings that will help you to sort out the contents of this bin. For example, one heading could be 'plastic'.

b) List each of the things in the bin under one of your headings. If one item doesn't fit you might need another heading.

at home...

You are going to investigate what you and your family throw out every week.

Keep a record of all the rubbish you and your family produce in a week.

a) Stick a sheet of paper on the wall in the kitchen. Hang a pen beside the sheet. At the top of the sheet, put the same headings as you used in Activity 1. Ask your family to help you. Each time they throw away something ask them to list it under the correct heading. At the end of the week compare your lists. Which is longest?

b) If possible, try to weigh the rubbish. (Better get permission before you use the bathroom scales!) Each time you fill your bin or a plastic sack, weigh it before you take it outside. Record the weight. Add the weights together at the end of the week.

Keep your results. You will need them when you get to page 105.

WHAT A LOAD OF RUBBISH!

→ Your home as a system

You can think of your home as a SYSTEM. A system has inputs and outputs. The inputs are all the RESOURCES that you use. The outputs are all the things you produce – including rubbish!

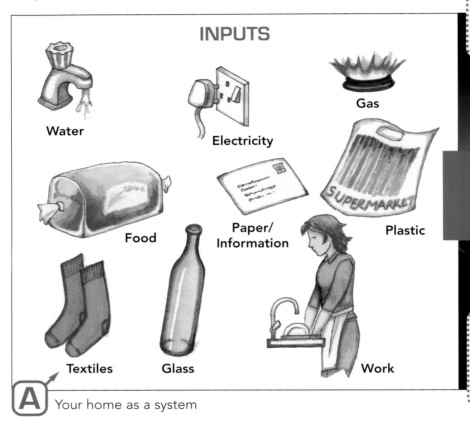

Even at night, when it looks like nothing is going on, there are still inputs coming into the house and outputs going out of it.

Many inputs into a home are invisible. Water, electricity and gas all arrive underground – and information can come through the air!

A Your home as a system

activity...

1 You are going to draw a diagram of your home as a system.
   a) First, think of the activities that happen in your home. For example, using the toilet! The drawings of inputs and outputs will give you other ideas. Think of at least five.
   b) Now, match each activity that you thought of with the inputs and outputs in **A**. For example, going to the toilet uses water and paper, and produces sewage.
   c) Draw a large diagram like the one on page 103 to fill a page. You might need to turn the page sideways. List all the inputs, activities and outputs in the three boxes. Leave some space between each one.

WHAT SHOULD WE DO WITH ALL OUR WASTE?

Some outputs are invisible too. You can't see energy, heat or knowledge!

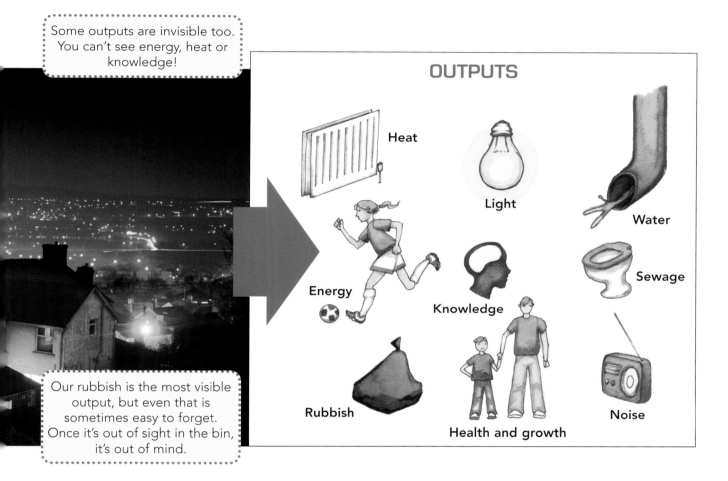

Our rubbish is the most visible output, but even that is sometimes easy to forget. Once it's out of sight in the bin, it's out of mind.

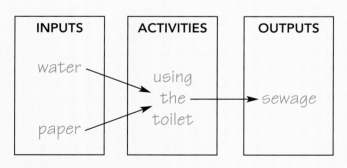

d) Draw arrows on your diagram to show the links between each activity and its inputs and outputs. An example is done for you. Some arrows may have to cross over each other. You could colour code each activity to make the diagram clear.

2 How could you improve the system to reduce the amount of rubbish, and other unnecessary outputs, your home produces? Suggest at least five ways to reduce, reuse or recycle in your home. For example:

Turn off the lights when they are not being used.

103

WHAT A LOAD OF RUBBISH!
➡ Where does the rubbish go?

Can you believe that each of us in the UK produces 400 kg of rubbish every year – the weight of six fully grown adults! That works out at about 23 kg per household, every week.

Graph **C** shows what is in all that rubbish.

B One person's rubbish for one year!

This waste is buried, burned or recycled.

- Most of it is buried in large holes in the ground called LANDFILL SITES.
- As we run out of space for new landfill sites, more waste is being burned in huge INCINERATORS.
- The alternative to landfill or incineration is RECYCLING. Recycling uses resources again. The government want us to recycle 30% of our waste by 2010. Diagram **D** shows that we've still got a long way to go!

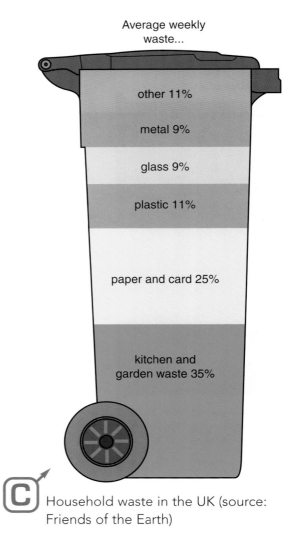

Average weekly waste...

- other 11%
- metal 9%
- glass 9%
- plastic 11%
- paper and card 25%
- kitchen and garden waste 35%

C Household waste in the UK (source: Friends of the Earth)

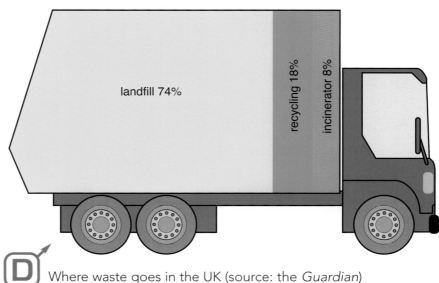

landfill 74% | recycling 18% | incinerator 8%

D Where waste goes in the UK (source: the *Guardian*)

activity...

1. Look at diagram **C**. Compare the waste from a typical UK household with your family's waste. You will need the results of your investigation on page 101 to do this.
   a) Did you sort your waste into the same types? Was kitchen and garden waste the main type of waste in your home?
   b) Do you produce more or less waste than the average (23 kg per week)?

2. Suggest reasons for any difference between your family and a typical household. (Clues: numbers of people, wealth, where you live.)

3. Look at chart **E**.
   a) How does the amount of waste recycled in the UK compare with the countries shown on the chart?
   b) Could we recycle more than we do? How could we do this?

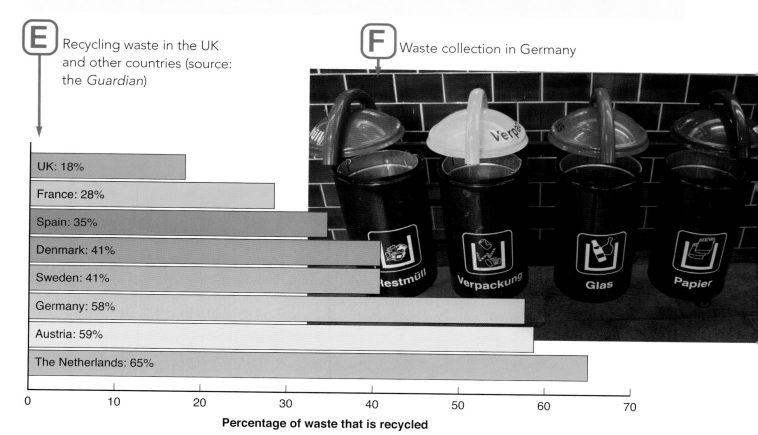

E Recycling waste in the UK and other countries (source: the *Guardian*)

- UK: 18%
- France: 28%
- Spain: 35%
- Denmark: 41%
- Sweden: 41%
- Germany: 58%
- Austria: 59%
- The Netherlands: 65%

Percentage of waste that is recycled

F Waste collection in Germany

make your own notes...

Over the next six pages you are going to study these three options for dealing with waste – landfill, incineration and recycling – in more detail.

As you study each option, make your own notes. To do this, divide a page in your book in half. On one side note the advantages. On the other side note the disadvantages. Use all the sources of information on each page, including photos, diagrams, maps and newspaper articles.

You could work in a group of three to do this, so that each person studies one spread. Later, you can share your notes. At the end you will use the notes to decide which is the best option to deal with our waste.

WHAT A LOAD OF RUBBISH!

→ Landfill – out of sight, out of mind

make your own notes...
Make notes about using landfill sites to deal with waste. Follow the instructions on page 105.

G A landfill site in the UK

Option 1: Landfill

Advantages	Disadvantages
Located away from urban areas where most people live.	Lorries have to transport waste long distances.

- Landfill sites are located away from urban areas where most people live
- Methane gas given off by waste adds to the problem of global warming
- Lorries have to transport waste long distances
- Landfill is one way to use huge holes in the ground created by mining
- The countryside is running out of space for new landfill sites
- Chemicals leak into the ground to pollute local rivers and water supply

WHAT SHOULD WE DO WITH ALL OUR WASTE?

H Landfill site – before, during and after use

Before

Mining companies dig opencast pits to extract resources like coal or gravel

When the resource has been extracted it leaves a huge hole in the ground

During

Rubbish is brought from urban areas to fill the hole

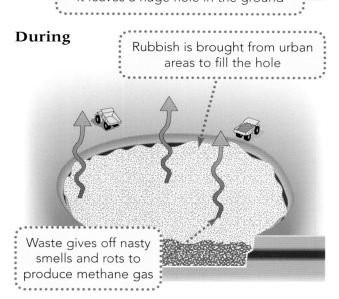

Waste gives off nasty smells and rots to produce methane gas

After

When it is full the whole site is covered with soil and grass to improve the landscape

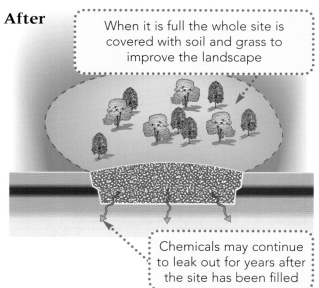

Chemicals may continue to leak out for years after the site has been filled

Villagers on the Brink
Paul Brown reports

The population of the ancient village of Boothorpe in Leicestershire faces ruin because of a large pit being dug next to their homes to take the county's household waste for 13 years.

The village was once in rolling countryside but now hangs on the edge of a cliff above an opencast mine which is being excavated. Once the mining is complete the company wants permission to fill it with rubbish.

Leicestershire has a mounting rubbish problem and will run out of landfill sites in two years.

Michelle Coombs, of the Council for the Protection of Rural England, said, 'There are so many landfill sites in south and central England it is difficult to get more than ten miles away from one. Local communities are affected by traffic noise and dust, the environment is polluted by chemicals and landscapes are damaged.'

Susan Reiblein, a Boothorpe resident, said, 'We have been robbed of the value of our homes because no one wants to buy next to a giant landfill.' Neighbour, Kulbir Kaur, claimed, 'The children cannot do their homework for the noise or play outside because of the dust. The whole thing is affecting our health. There is no future here for us, but we cannot move.'

I Article adapted from the *Guardian*

activity...

Imagine that you are one of the residents in Boothorpe. Write a letter to the government to express your feelings about the landfill site. Tell them what *you* think should happen to the rubbish. Here are some words and phrases you could use in the letter.

pollution   health
chemicals   landscape   property
I believe...   Obviously...
It is clear that...   Therefore...

107

WHAT A LOAD OF RUBBISH!

→ Incineration – going up in smoke

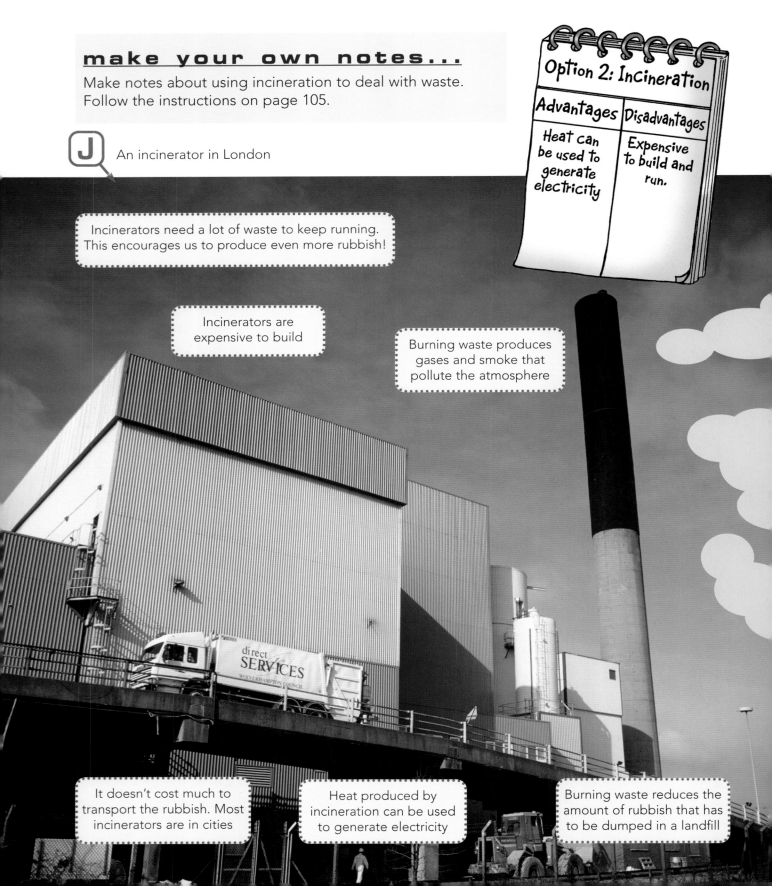

make your own notes...
Make notes about using incineration to deal with waste. Follow the instructions on page 105.

Option 2: Incineration

Advantages	Disadvantages
Heat can be used to generate electricity	Expensive to build and run.

J An incinerator in London

Incinerators need a lot of waste to keep running. This encourages us to produce even more rubbish!

Incinerators are expensive to build

Burning waste produces gases and smoke that pollute the atmosphere

It doesn't cost much to transport the rubbish. Most incinerators are in cities

Heat produced by incineration can be used to generate electricity

Burning waste reduces the amount of rubbish that has to be dumped in a landfill

There are thirteen incinerators for household waste in Britain, with another five being built. You can see their locations in map **K**. The government wants to build more incinerators to deal with the growing amount of waste. This could lead to more of the POLLUTION problems you can see in **L**.

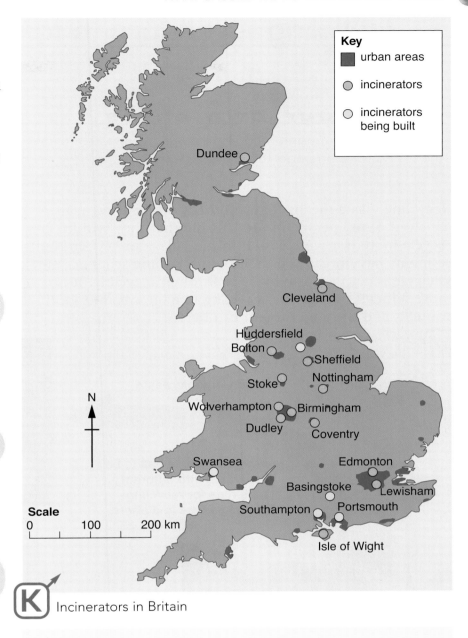

Incinerators in Britain

Carbon dioxide (CO_2)
Burning waste produces CO_2. The amount of CO_2 in the atmosphere is growing and is the main cause of global warming.

Nitrogen oxides (e.g. NO_2), sulphur dioxide (SO_2)
These gases mix with rainwater to cause acid rain that kills trees and fish.

Particulates (soot and ash)
These are tiny particles that create dirt and affect people with breathing problems. Asthma in cities is a growing problem.

Chemicals (e.g. dioxins)
Incinerators burn all sorts of things that produce harmful chemicals. Some plastics produce dioxins that cause cancer.

L Types of air pollution from incinerators

activity...

1 Look at map **K**.
   a) Which is the nearest incinerator to where you live? How far away is it? (Use the scale on the map.)
   b) Would you like to have an incinerator close to your home? Explain why.

aim high...

2 You are the government's energy minister.
   a) Choose the best location for a new incinerator in Britain. Where would you locate it on the map?
   b) Write a short report (one or two paragraphs) to justify your choice.

WHAT A LOAD OF RUBBISH!

→ Recycling – resources for ever

make your own notes...
Make notes about using recycling to deal with waste. Follow the instructions on page 105.

Recycling waste means that resources can be used again. Nine out of ten people in the UK said that they would recycle more if it were easier. In the village of Wye in Kent residents have organised their own recycling scheme. The village has cut its waste by 75% through recycling. It is more difficult to organise recycling in a big city, but local councils and the government can help.

 The WyeCycle – a recycling scheme in one village

non-organic waste

non-organic waste has to be separated into different materials for recycling. This takes time and effort

waste collection

organic waste (food and plant material) is collected separately from non-organic waste

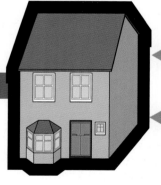

organic waste is turned into compost

WHAT SHOULD WE DO WITH ALL OUR WASTE?

Household

I hate chucking things out – it seems such a waste. But what am I supposed to do? Everything I buy at the supermarket is wrapped in about three layers of plastic.

Supermarket

Everybody expects food to be fresh. That's why we wrap it properly. They even expect us to provide bags for them to take their shopping home in.

Government

We are building more incinerators. We would rather that people recycle their waste, but it takes time to change habits and we have a problem of what to do with the waste now.

 Excuses, excuses!

metal
textiles
glass
plastic
paper

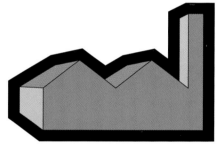

materials sent to factories to be used again, but they need to be transported by lorries

people can buy recycled products in the shops

food is delivered direct from local farms – there is no wasteful packaging

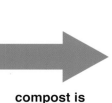

compost is used to enrich soil and to grow food

activity...

1 Look at the drawing of the WyeCycle in **M**. Describe how each of these types of waste is recycled:
   a) non-organic waste like plastic, glass, paper, etc.
   b) organic waste.

discuss...

2 Read the excuses from households, supermarkets and government in **N**.
   a) For each group, do you think it is a good excuse or not? Give a reason.
   b) Suggest one thing that each group could do to help to make recycling work.

111

WHAT A LOAD OF RUBBISH!

→ Share your notes

activity...

You are going to use the notes you have made about the three main options: landfill, incineration and recycling. If you worked in a group of three, this is the time to share your notes.

1 Read the statements in the box below. Without looking at your notes, test yourself on what you have learnt. Which of the three options is each statement about – landfill, incineration or recycling? Is the statement an advantage or a disadvantage?

2 Check your notes. Have you included all these statements in your notes? If not, you could add them now. Make sure they go in the correct column. (You may have made more notes than this. That's fine!)

- Lorries transport waste over long distances
- Villages can be ruined by noise and dust
- Chemicals leak into the ground
- Most people would like to do it
- They are expensive to build
- Groups of people can organise it together
- Different types of waste need to be separated
- Waste does not need to be transported far
- They are located away from large urban areas
- Heat can be used to produce electricity
- Compost helps to enrich the soil
- They need a lot of waste to keep running
- The countryside is running out of space
- They use large holes in the ground
- They pollute the atmosphere with harmful gases
- It saves using more resources
- They can deal with waste quickly – now
- It is more difficult to organise in a big city

112

your final task...

Now, it is time to decide: What should we do with all our waste?

1 Score the points you have made in your notes. Give a positive (+) score for an advantage and a negative score (–) for a disadvantage. Then decide how important each point is:
- + or – 3 if it is very important
- + or – 2 if it is quite important
- + or – 1 if it is not very important.

Write the score beside each point in your notes.

2 Work out a total score for each option. Add the positive scores and subtract the negative scores. (For example, +15 and –8 = +7.) Which option gets the highest score? Now, make your decision.

3 Write a letter to your local council to tell them what you think we should do with all our waste. Here is a writing frame you could use to get you started.

Wasteham Comprehensive School

Landfill Lane
Wasteham WA1 1AA
1 May 2006

Dear Sir/Madam,

I have been studying what we do with our waste in geography lessons at school. We looked at three ways of dealing with waste – landfill sites, incineration and recycling. At the end of the lessons we had to decide what was the best way to deal with our waste.

In my opinion, the best way to deal with waste is: ...
The main advantages are: ...
I do not think it is so good to ...
The main disadvantages are: ...
The worst way to deal with waste is ...
The main disadvantages are: ...

I hope that you will consider my opinion as you decide what the council should do with our waste in the future.

Yours faithfully,
Ella Davis

8 Look again at the United Kingdom

What would a newcomer to the UK want to know?

KEY CONCEPT
- Place
- Diversity
- Space

Every year about 2500 children and teenagers come to the UK alone as refugees. They are fleeing war, persecution or poverty in their own countries. You may be one of them. But, if not, imagine that you have just come to live in the UK. All you know about it is what you have seen on TV. You don't know anybody here. You don't even speak the language very well. What would you want to know about your new home?

I WANT TO KNOW...

... how to speak English

... what is the difference between England and the UK

... where the Queen lives

... what the English eat

... how to find a nice place to live

... when I can vote in a General Election

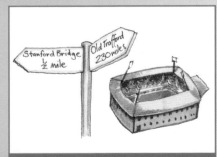

... directions to the nearest Premier League football ground

WHAT WOULD A NEWCOMER TO THE UK WANT TO KNOW?

coming up...

Through this unit you will find out a lot about the UK. Some of it may be new to you, even if you have lived here all your life.

your final task...

You will use this information and your own knowledge to make a PowerPoint presentation about the UK for newcomers to this country.

You must include slides covering at least three of the topics you have studied in this unit. Which topics you choose is up to you.

starter...

1. Look at all the possible things a newcomer might want to know about the UK. Which do you think would be *most* important for a newcomer to know? Which *least* important? Give your reasons.
2. You may think something important is missing. If so, add it to the list and explain why it is important.
3. Which of these things do you think have anything to do with geography? Explain why.

... what jobs I could do when I'm older

... what the weather will be like where I live

... the words of the National Anthem

... why I get called names on the street

... what subjects I will study at school – and why they are useful

LOOK AGAIN AT THE UNITED KINGDOM
➜ Four countries in one

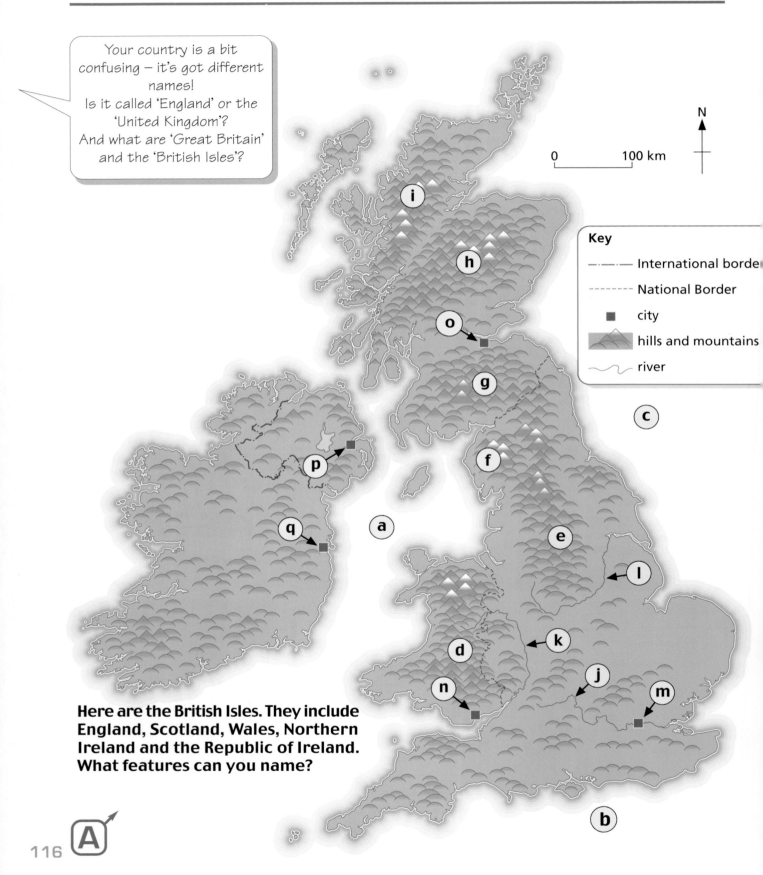

Your country is a bit confusing – it's got different names!
Is it called 'England' or the 'United Kingdom'?
And what are 'Great Britain' and the 'British Isles'?

Key
- —·—·— International border
- - - - - - National Border
- ■ city
- ⛰ hills and mountains
- ～ river

Here are the British Isles. They include England, Scotland, Wales, Northern Ireland and the Republic of Ireland. What features can you name?

A

WHAT WOULD A NEWCOMER TO THE UK WANT TO KNOW?

activity...

1 Look at the shapes below. Find them on map **A**.
   a) Name each country.

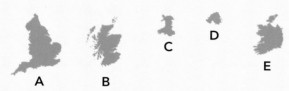

   b) Match each country with one of these flags:

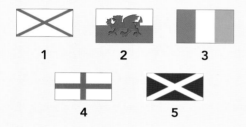

2 You are going to draw a sketch map of the United Kingdom.
   a) First draw an outline of the British Isles like the one here.

   b) Add the country borders to your map.
   c) Shade in the countries that are in the UK.
   d) Label the countries in the UK.
3 Name the seas, mountains, rivers and cities marked on map **A**, for example, *a = Irish Sea*. You can check on the map on page 134.

Sort out the confusion!

- **The British Isles** is made up of two large islands – Great Britain and Ireland.
- **Great Britain** is the largest island in the British Isles. It includes England, Scotland and Wales. Ireland includes Northern Ireland and the Republic of Ireland.
- **The United Kingdom** is made up of four countries – England, Scotland, Wales and Northern Ireland.
  The Republic of Ireland is a separate country.

 The Union Jack combines the flags of England, Northern Ireland and Scotland.

look again...

Look again at these two pages. Why might this information be useful to a newcomer to the UK? What might they want to know? Here is how you could summarise the information. At the end of the unit you will decide whether to include this in your presentation.

WHAT IS THE UK?

- The United Kingdom is made up of four countries – England, Scotland, Wales and Northern Ireland.
- The United Kingdom is not the same as the British Isles or Great Britain.

LOOK AGAIN AT THE UNITED KINGDOM

➡ You are not the first

> One reason I wanted to come to the UK is because it has always been one of the safest places for immigrants.

If each of us could trace our family far enough back we would find that our ancestors were immigrants. Diagram **B** shows the main groups of immigrants that have arrived over the centuries. Do you know which group of immigrants you are descended from?

B The main groups of immigrants to the UK – with the date of their first arrival, their reason for coming and what they did when they arrived

Irish 1840
Came to escape poverty and famine in Ireland. They helped to build cities and worked in factories.

Celts 800 BC
Came from Central Europe to find land to farm. They lived in villages.

Vikings AD 800
Came from Denmark and Norway to raid settlements along the coast. Later they settled here themselves and ruled part of the country.

West Indians 1950
Invited by the UK government to work here when there was a shortage of workers after Second World War.

Romans AD 43
Came from Rome (in Italy) to conquer. They brought people from all over the Roman Empire who settled here. The Romans built the first cities and our first road network.

Africans 1980
Came to escape war, poverty and persecution in their countries. Many came from English-speaking countries, once part of the British Empire.

Saxons AD 500
Came from northern Europe to farm. They built settlements and took over the country. The English language comes originally from the Saxons.

Normans 1066
Came from northern France to conquer England and Wales. They settled here and built castles to defend their settlements.

Indians, Pakistanis, Bangladeshis 1955
Invited to fill important jobs, especially in transport, health and textile industries.

Jews 1880
Came mainly from Eastern Europe to escape religious persecution that continued in the twentieth century.

Eastern Europeans 1990
Came after the fall of communist governments in their countries. They came for work, but some also came to escape war and persecution.

WHAT WOULD A NEWCOMER TO THE UK WANT TO KNOW?

Here are three types of immigrant:

- INVADERS – people who come to take over the country by force.
- ECONOMIC MIGRANTS – people who come to work and to improve their quality of life. A country sometimes invites economic migrants if it needs more workers.
- REFUGEES – people who are forced to leave their own country through fear of war or persecution.

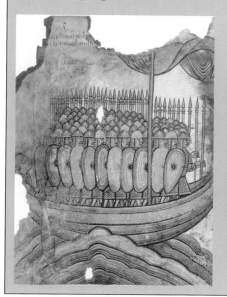

discuss...

When asked why they chose to come to the UK these are the main reasons that refugees give:

- the UK is a safe, tolerant, democratic country
- they have friends or relatives living here
- the UK is a well-known country with many famous people
- there is plenty of work in the UK
- English is a widely spoken language
- their country has historic links with the UK.

1 a) If you had to leave the UK, which country would you go to?
b) Why would you choose that country? Give reasons.

activity...

2 Look at diagram **B**.
You are going to draw a timeline to show the main groups of immigrants that have come to the UK.
a) Draw a line, like the one below, across a page in your book.
b) Label the immigrant groups above or below the line with an arrow pointing to the date of their first arrival.

c) Underline each group in one of three colours to show if they were invaders, economic migrants or refugees. Some groups may have two colours.
d) Write two sentences to describe the pattern you can see on the line. How has the number of new groups changed? How has the type of immigrant changed?

look again...

Look again at these two pages. Write one or two key points to put on a slide. Later you will decide whether to use these in your presentation.

MULTICULTURAL BRITAIN
-
-

LOOK AGAIN AT THE UNITED KINGDOM

→ Too crowded?

People say that the UK is already a crowded country, so there's no room for people like me. But is that really true?

Map **C** is a POPULATION DENSITY map of the UK. It shows where people live. The dark colours show parts that are crowded – lots of people live there. The light colours show empty parts – where few people live.

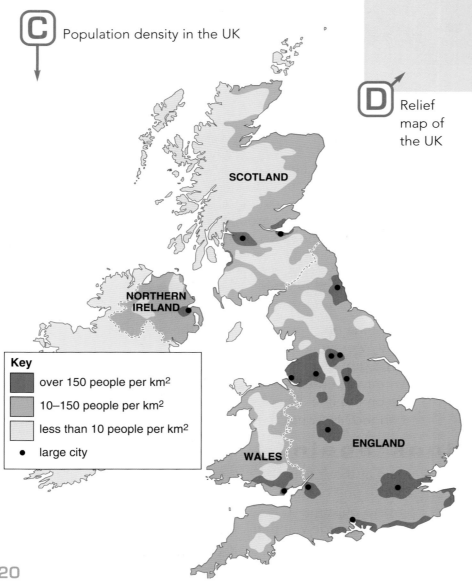

C Population density in the UK

D Relief map of the UK

Key: high ground

Key:
- over 150 people per km²
- 10–150 people per km²
- less than 10 people per km²
- • large city

activity...

1. Look at Map **C**. Find where you live.
   a) What is the population density where you live?
   b) Do you think your area has too many people, not enough people or just right? Give reasons.

2. There is a strong link between relief (the height and shape of the land) and population density. See if you can work out how they are linked. Map **D** is a relief map of the UK. Compare the pattern on maps **C** and **D**. Write two sentences starting like this:

   Areas with high population density are found ...

   Areas with low population density are found ...

WHAT WOULD A NEWCOMER TO THE UK WANT TO KNOW?

  Most of Scotland has LOW POPULATION DENSITY. Some remote areas of Scotland have only one or two people per square kilometre. That could get lonely but some people love it!

F Cities like London have HIGH POPULATION DENSITY. London has over five thousand people per square kilometre. But there is still room to move

discuss...

3 Compare photos **E** and **F**. Where would you most like to live – is it an empty place or a crowded place? Give reasons.

4 Where would you advise a newcomer to live? Give reasons.

aim high...

5 Explain the link between relief and population density. The photos will give you some clues. For example:

Mountain areas have low population density because the land is too steep to build on.

look again...

Look back at these two pages. Write one or two key points to put on a slide. Later you will decide whether to use these in your presentation.

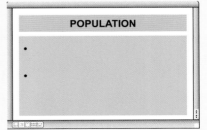

LOOK AGAIN AT THE UNITED KINGDOM

→ Always raining?

The weather in the UK is so unpredictable. How can I know what the weather is going to be like where I live?

We are surrounded by weather forecasts: on TV, the radio, the internet and in newspapers. They tell us what the weather will be like for the next few days. If you look carefully at map **G** you may be able to see a pattern in the weather.

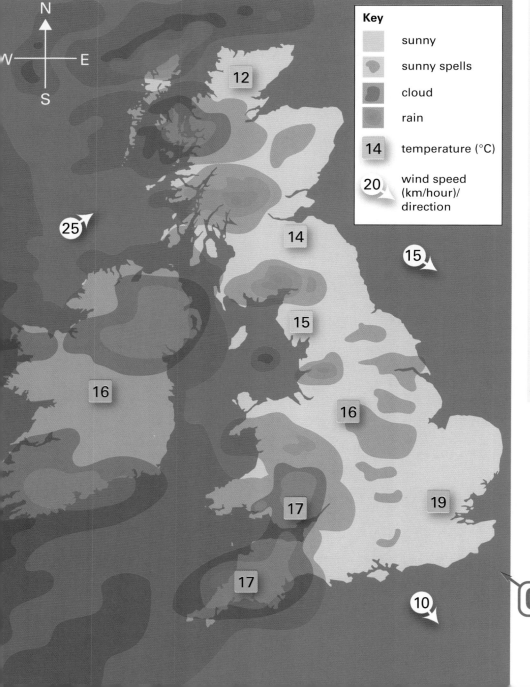

G Typical weather for the British Isles for one day in June

activity a...

1 Find where you live on map **G**. Describe what the weather is like in your area. Mention the temperature, rainfall, sunshine/cloud and wind.

2 Write a short script for the TV weather forecast, based on the weather shown on map **G**. Mention each part of the British Isles in your script.

Eastern areas will be...

Further west it is likely to...

Highest temperatures will be...

The worst weather is expected...

WHAT WOULD A NEWCOMER TO THE UK WANT TO KNOW?

Do you remember the difference between weather and climate (see page 42)?
Weather is what happens from day to day. Map **G** shows weather. **Climate** is the average weather over many years. The maps in **H** show climate.

1 Average summer temperature (July)

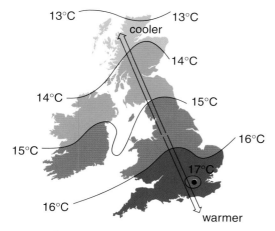

2 Average winter temperature (January)

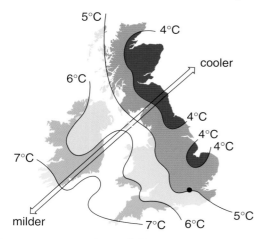

3 Average annual rainfall

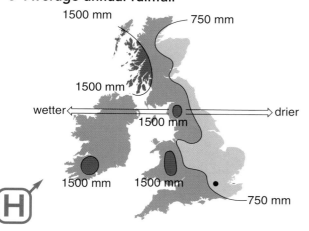

activity b...

3 Find London on the maps in **H**. Work out for London:
   a) the average summer temperature
   b) the average winter temperature
   c) the annual rainfall.

4 a) Draw a map of the British Isles divided into four areas, like this:

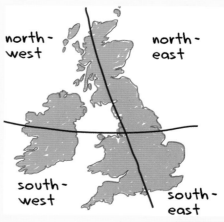

b) Describe the climate in one of the four areas on your map. Mention summer and winter temperatures and annual rainfall in each area. For example:

In the north-west summers are cool. The temperature is 13 to 14 °C. The winters are mild with a temperature of 5 to 6 °C. The annual rainfall is high. It is over 1500 mm.

look again...

Look back at these two pages. Write one or two key points for your slide. Later you will decide whether to use these in your presentation.

LOOK AGAIN AT THE UNITED KINGDOM

→ Nice or nasty?

> When I first arrived in the UK I was sent to live in a small village. It was beautiful, but there was no one else like me there.

Look at the photos. The photos in **I** show a **rural** environment in the UK. The photos in **J** show an **urban** environment. Where would you rather live and why?

I Rural environments

activity a...

1 a) Make an environmental quality evaluation chart like the one below. This chart helps you judge the quality of a particular environment.

← High quality	5	4	3	2	1	Low quality →
Attractive						Ugly
Interesting						Boring
Clean						Dirty
Safe						Dangerous

Each row has a pair of opposite words. When you fill in the chart you will put a tick (✓) somewhere between them. For example, for row one:
- tick (✓) 5 if the environment is attractive
- tick (✓) 4 if the environment is quite attractive
- tick (✓) 3 if the environment is neither attractive nor ugly
- tick (✓) 2 if the environment is quite ugly
- tick (✓) 1 if the environment is ugly.

WHAT WOULD A NEWCOMER TO THE UK WANT TO KNOW?

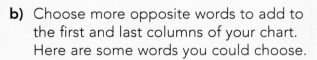
Urban environments

activity b...

2 Fill out two charts – one for a rural environment using the photos in **I**, one for an urban environment using the photos in **J**.

3 Work out a total score for each environment. Add up the number of ticks, times by the value of the column. (For example, 10 ticks in the 5 column would be a score of 50.)

aim high...

4 What don't the photos tell you? What else would you need to know to decide if this was a good place to live? List some other questions you would ask before you make a decision.

b) Choose more opposite words to add to the first and last columns of your chart. Here are some words you could choose.

empty/crowded   noisy/quiet   modern/old
natural/built-up   many cars/few cars
many shops/few shops

c) Decide which word from each pair describes a 'high quality' or 'low quality' environment. This is your opinion. There is no right answer. Some people think empty is good, others think crowded is good. Write the words in the correct column.

look again...

Look back at these two pages. Write one or two key points for your slide. Later you will decide whether to use these in your presentation.

ENVIRONMENT
•
•

125

LOOK AGAIN AT THE UNITED KINGDOM

→ The best jobs

> Some people think that refugees come to the UK for any easy life – to live off benefit. That's not true! I want a good job. That's the way to get ahead.

In geography, jobs are sometimes called ECONOMIC ACTIVITIES. That is because you are paid to do them. Bad luck – being a school pupil is not an economic activity!

There are three main types of economic activity. Graph **K** shows how economic activities in the UK have changed.

K Changing economic activities in the UK

> **PRIMARY ACTIVITIES**
> These extract or grow natural resources from the land or sea. Farming, fishing and mining are primary activities.

> **SECONDARY ACTIVITIES**
> These make or manufacture things from natural resources, or assemble things that have already been made. Making steel from iron ore and assembling cars are secondary activities.

Farming in the nineteenth century – a primary activity

Factory work in the twentieth century – a secondary activity

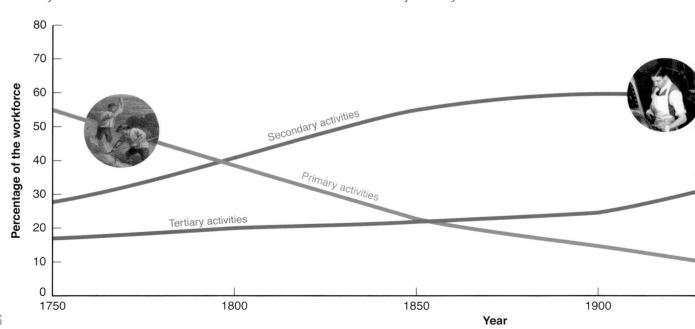

WHAT WOULD A NEWCOMER TO THE UK WANT TO KNOW?

activity...

1 Read the list of jobs in the box below. Classify them into three groups – primary, secondary and tertiary activities. List them under the correct headings.

> coal miner   builder
> pop singer   firefighter   bus driver   fish farmer
> nurse   oil rig worker   police officer   car mechanic
> shop assistant   football player   gardener
> bank manager   television factory worker
> computer programmer   teacher

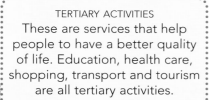

TERTIARY ACTIVITIES
These are services that help people to have a better quality of life. Education, health care, shopping, transport and tourism are all tertiary activities.

Education in the twenty-first century – a tertiary activity

2 a) Draw a large table, like the one below, to show how economic activities in the UK have changed.
  b) Look at graph **K**. Find the percentage of the workforce in primary, secondary and tertiary activities in 1750. Check the figures in the table below. Notice that the figures add up to 100.
  c) Now, find the percentage figures for the other dates on the table. Write them in your table. For each date the figures should add up to 100.

	1750	1800	1850	1900	1950	2000
Primary	55					
Secondary	28					
Tertiary	17					

3 a) Which of the jobs in the list could someone in your area do?
  b) Which of the jobs would you most like to do?
  c) Which of them needs the most education?

look again...

Write one or two key points for your slide. Later you will decide whether to use these in your presentation.

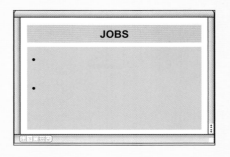

LOOK AGAIN AT THE UNITED KINGDOM

➜ Have your say

> I had to leave my country because it had a bad government. You are lucky to live in a country with a democratic government. How does it work?

In the UK DEMOCRACY means that everyone (over 18) can vote to choose the government. Many refugees come from countries with no democracy. That is why there is often war and persecution there.

Democracy in action

When you are 18 you will be allowed to vote in General Elections so it will be you who decides who should govern the country. You will elect an MP who will go to Parliament and have to vote on any new laws.

activity...

1 You may have to wait until you are 18 to vote in a real election but you don't have to wait until then to vote in a mock-election. Imagine it is election time. The four candidates opposite want you to vote for them. Who will you vote for and why?
   a) Choose the candidate whose ideas you most like.
   b) Vote in a secret ballot for the candidate you chose in a). You can vote for only one person.
   c) Your teacher will announce the class result.

discuss...

2 Talk about these questions with a partner or in a small group.
   a) Do you think we are lucky to live in a democracy? Why?
   b) Should people under 18 be allowed to vote? Why?
   c) Is there any democracy in your school, for example, a school council? If so, is it effective? If not, do you think it would be a good idea?

There are over 60 million people in the UK. The national government is in London.

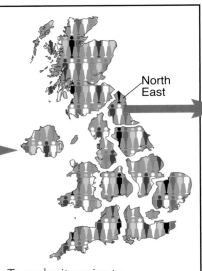

To make it easier to govern, the UK is divided into REGIONS. Scotland, Wales and Northern Ireland have their own regional governments.

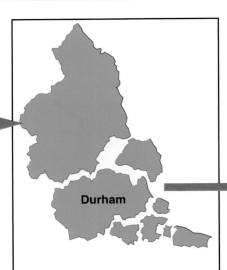

Regions are divided again into counties. Each COUNTY has its own local government responsible for things like schools and police.

WHAT WOULD A NEWCOMER TO THE UK WANT TO KNOW?

Candidate 1
I promise to end all your problems.
Our biggest local problem is crime. Crime is ruining this community.
I suggest we end crime in this area by sending all offenders to prison however small their crime.
Now on to housing. There are not enough houses round here for people like you. We need more affordable houses. I propose selling off all school playing fields so that we can build cheap housing that everyone can afford.

Candidate 2
If you elect me I will work hard for you.
I will do all I can to help you whether your problem is that you are poor and can't afford to keep warm in winter or live next to a noisy neighbour and want help keeping them quiet.
On the national scene – my biggest priority will be protesting against war. Wars are expensive and unnecessary. People are killed and maimed. There are better ways to solve problems. I will always vote against wars.

Candidate 3
If you elect me you elect a global citizen.
Our local problems are tiny compared to the global problems of pollution and climate change.
We must cut the use of the car. Petrol fumes choke our air and cause global warming.
I propose closing the town centre to traffic and pedestrianising all the streets. We will provide cheap and efficient public transport so people do not need cars and we will stop people flying in aeroplanes.

Candidate 4
I believe in fairness.
This community, this world is divided into rich and poor. The rich are getting richer and the poor are getting poorer.
I propose increasing taxes on rich people and giving more money to poor people so they can afford better lives.
I propose improving the hospitals and health service so that everyone gets the best care available. If that means spending more money that is OK. Let's make Britain healthier.

aim high...

3 You might think that the speeches are not very good ones. Choose an issue that you really care about and write your own mini-speech explaining to the rest of the class what you would try to do about it if you were elected an MP.

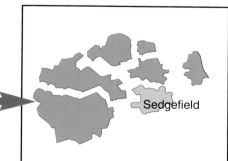

Constituencies are even smaller areas. Each CONSTITUENCY elects an MP (Member of Parliament) to represent them in Parliament. The MP for Sedgefield was once Tony Blair. There are 646 constituencies in the UK.

look again...

Look back at these two pages. Write one or two key points for your slide. Later you will decide whether to use these in your presentation.

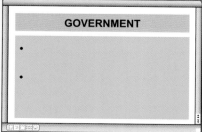

LOOK AGAIN AT THE UNITED KINGDOM

your final task...

You are going to make a PowerPoint presentation about the United Kingdom for people arriving in the country for the first time. The government want immigrants to understand more about the UK to help them to become citizens in the country. Your presentation is important because first impressions can last a long time.

You have probably used PowerPoint before. With PowerPoint you create a sequence of slides on the computer screen. You can show your presentation to an audience using a data projector or interactive whiteboard. If you haven't got PowerPoint, a projector or a whiteboard, don't worry! The most important thing is your ideas. You could present the same ideas in a booklet for new arrivals to read.

This is what you have to do:

1 Think about your audience
What do newcomers to the UK really want to know? What would really surprise them about the UK?

2 Plan your presentation
Bring together the main ideas from this unit. Each section looked at one main idea, for example, 'Weather and climate'. For each idea you wrote one or two key points.

You need to decide:
a) Which of the topics you think are relevant to your audience – you must include at least three topics from the unit.
b) What else you would like to include in your presentation – there are lots of ideas from the rest of the book. All of the case studies have been about the UK.

3 Make your slides
You should make a slide for each of the main ideas that you studied. Here is an example:

Title to describe the main idea. You can make up your own title if you prefer.

WHAT IS THE UNITED KINGDOM?

- The United Kingdom is made up of four countries – England, Scotland, Wales and Northern Ireland.
- The United Kingdom is not the same as the British Isles or Great Britain.

Key points – keep them short for your presentation.

Graphic – you can import maps, graphs or cartoons which your teacher will give you or draw your own.

WHAT WOULD A NEWCOMER TO THE UK WANT TO KNOW?

You can vary the layout of your slides to make your presentation more interesting. Here are some of the layouts that you could create with PowerPoint. There are lots more.

A title slide

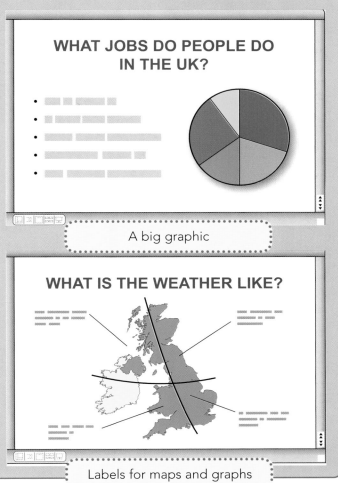

A big graphic

Labels for maps and graphs

4 Make your presentation to your class

As you show each slide, don't just read out what the slide says. Add extra details as you talk, or allow your audience to ask you questions.
For example:

Did you know that we all started out as immigrants? My granny came from France. She married an Englishman. Their son got together with an Irish woman and they produced me! So I'm half Irish, one-quarter French and one-quarter English. But I was born here, so I'm British.

Coverage of Key Concepts

■ Primary Key Concept ■ Secondary Key Concept

Book 1	Place	Space	Scale	Interdependence	Environmental interaction and sustainable development	Physical and human processes	Cultural understanding and diversity
1 Your place and mine	School, as experienced by a disabled pupil	Disabled access within a school	Personal				Values and attitudes towards disability
2 My spaces	Local area	Links between a place and its surroundings at a range of scales	Personal, local, wider local, national & global	Social and economic links between a place and the wider local area, the UK and the world			Personal connections with people and places in other parts of the world
3 Survivor!	An imaginary island, based on a small Hebridean island	Location of water, shelter and food on the island	Small island	Appreciate the difficulties people face when they are isolated	Human dependence on the natural environment for water, shelter and food	Water cycle Factors affecting the weather Ecosystems	
4 City – past, present, future	Manchester	Location of Manchester, changes in the city over time and implications for people	City	Movement of people and goods to and from Manchester	How the natural environment has turned into an urban environment	Population growth Urbanisation Regeneration	Multicultural Manchester – the contribution of different groups to the city
5 Shop until you drop!	Villlage in the Lake District	Changes in the distribution of shops and services	Village and surrounding area	Impact of new supermarket on smaller shops			Appreciate how changes in shopping affect communities
6 Flood disaster	Tewkesbury and the Rivers Severn and Avon	Location of Tewkesbury and pattern of flooding in and around the town	Town and river valley		Impact of flooding on people Ways in which human actions affect the risk of flooding	Flooding	
7 What a load of rubbish	Local examples within the UK	Distribution of landfill sites and incinerators in UK	Links between personal and national	Disposal of domestic waste in other places	Impact of waste disposal on the environment. Recycling as a sustainable option	Recycling	Values and attitudes towards waste disposal and recycling
8 Look again at the United Kingdom	UK	Location of places in the UK Distribution of population and climate in the UK	National	Migration to the UK from other parts of the world	Relationship between physical environment and quality of life	Migration	Values and attitudes towards immigration Comparing quality of life in rural and urban areas

Reference map of the UK

→ Ordnance Survey 1:50,000 map symbols

Communications

ROADS AND PATHS
VOIES DE COMMUNICATION
STRASSEN UND WEGE

Not necessarily rights of way

- Service area
- Junction number
- M 1 Motorway (dual carriageway) / Autoroute (chaussées séparées) avec aire de service et échangeur numéroté / Autobahn (zweibahnig) mit Servicestation und Anschlusstelle sowie Nummer der Anschlusstelle
- Elevated / En Viaduc / Erhöht
- Motorway under construction / Autoroute en construction / Autobahn im Bau
- A 470 Primary Route / Itinéraire principal / Fernstrasse
- Primary route under construction / Itinéraire principal en construction / Fernstrasse im Bau
- Dual carriageway / Chaussées séparées / Zweibahnige Strasse
- A 493 Main road / Route principale / Hauptstrasse
- Main road under construction / Route principale en construction / Hauptstrasse im Bau
- B 4518 Secondary road / Route secondaire / Nebenstrasse
- Narrow road with passing places / Route étroite avec voies de dépassement / Enge Strasse mit Ausweichstelle
- B 885 Road generally more than 4m wide / Route généralement de plus de 4m de largeur / Strasse, im allg. über 4m breit
- Road generally less than 4m wide / Route généralement de moins de 4m de largeur / Strasse, im allg. unter 4m breit
- A 855 Other road, drive or track / Autre route, allée ou sentier / Sonstige Strasse, Zufahrt oder Feldweg
- Path / Sentier / Fussweg
- Gradient: steeper than 20% (1 in 5) / Pente: Supérieure à 20% (1 pour 5) / Steigung über 20%
- 14% to 20% (1 in 7 to 1 in 5) / 14% à 20% (1 pour 7 à 1 pour 5) / 14% bis 20%
- Unfenced / Sans clôture / Nicht eingezäunt
- Footbridge / Passerelle / Fussgängerbrücke
- Bridge / Pont / Brücke
- Gates / Barrières / Schranken
- Road tunnel / Tunnel routier / Strassentunnel
- Ferry (passenger) / Bac pour piétons / Personenfähre
- Ferry (vehicle) / Bac pour véhicules / Autofähre
- Ferry P
- Ferry V

PRIMARY ROUTES
These form a network of recommended through routes which complement the motorway system

PUBLIC RIGHTS OF WAY
DROIT DE PASSAGE PUBLIC
ÖFFENTLICHE WEGERECHTE

- Footpath
- Road used as a public path
- Bridleway
- Byway open to all traffic

Public rights of way shown on this map have been taken from local authority definitive maps and later amendments. The map includes changes notified to Ordnance Survey by 1st August 1997. The symbols show the defined route so far as the scale of mapping will allow.

Rights of way are not shown on maps of Scotland

Rights of way are liable to change and may not be clearly defined on the ground. Please check with the relevant local authority for the latest information

The representation on this map of any other road, track or path is no evidence of the existence of a right of way

OTHER PUBLIC ACCESS
AUTRES ACCES PUBLICS
ANDERE ÖFFENTLICHE WEGE

- • • • Other route with public access {not normally shown in urban areas}

The exact nature of the rights on these routes and the existence of any restrictions may be checked with the local highway authority. Alignments are based on the best information available. These routes are not shown on maps of Scotland

- ● ● National/Regional Cycle Network
- — — Surfaced cycle route
- [4] National Cycle Network number
- [8] Regional Cycle Network number
- ♦ National Trail, European Long Distance Path, Long Distance Route, selected Recreational Routes

Danger Area — Firing and Test Ranges in the area. Danger! Observe warning notices. Champs de tir et d'essai. Danger! Se conformer aux avertissements. Schiess und Erprobungsgelände. Gefahr! Warnschilder beachten.

RAILWAYS
CHEMINS DE FER
EISENBAHNEN

- Track multiple or single
- Track under construction
- Light rapid transit system, narrow gauge or tramway
- Bridges, Footbridge
- Tunnel
- Station, (a) principal
- Siding
- Light rapid transit system station
- LC Level crossing
- Viaduct

General Information

LAND FEATURES

- Electricity transmission line (pylons shown at standard spacing)
- Pipe line (arrow indicates direction of flow)
- Buildings
- Public building (selected)
- Bus or coach station
- Place of Worship { with tower / with spire, minaret or dome / without such additions }
- ° Chimney or tower
- Glass Structure
- H Heliport
- △ Triangulation pillar
- Mast
- Wind pump/wind generator
- Windmill with or without sails
- + Graticule intersection at 5' intervals
- Cutting, embankment
- Quarry
- Spoil heap, refuse tip or dump
- Coniferous wood
- Non-coniferous wood
- Mixed wood
- Orchard
- Park or ornamental ground
- Forestry Commission access land
- National Trust-always open
- National Trust-limited access, observe local signs
- National Trust for Scotland

ORDNANCE SURVEY SYMBOLS

Tourist Information

TOURIST INFORMATION | RENSEIGNEMENTS TOURISTIQUES / TOURISTENINFORMATION

⚑	Camp site	Terrain de camping / Campingplatz
🚐	Caravan site	Terrain pour caravanes / Wohnwagenplatz
	Garden	Jardin / Garten
	Golf course or links	Terrain de golf / Golfplatz
i	Information centre, all year / seasonal	Office de tourisme, ouvert toute l'année / en saison / Informationsbüro, ganzjährig / saisonal
	Nature reserve	Réserve naturelle / Naturschutzgebiet
P&R	Parking / Park and ride, all year / seasonal	Parking / Parking et navette, ouvert toute l'année / en saison / Parkplatz / Park & Ride, ganzjährig / saisonal
✕	Picnic site	Emplacement de pique-nique / Picknickplatz
PC	Public convenience (in rural areas)	Toilettes (à la campagne) / Öffentliche Toilette (in ländlichen Gebieten)
	Selected places of tourist interest	Endroits d'un intérêt touristique particulier / Ausgewählter Platz von touristischem Interesse
✆	Telephone, public / motoring organisation	Téléphone, public / associations automobiles / Telefon, öffentlich / automobilklub
	Viewpoint	Point de vue / Aussichtspunkt
V	Visitor centre	Centre pour visiteurs / Besucherzentrum
	Walks / Trails	Promenades / Wanderwege
▲	Youth hostel	Auberge de jeunesse / Jugendherberge

Technical Information

NORTH POINTS

Difference of true north from grid north at sheet corners

NW corner	NE corner
1° 03' (19 mils) E	0° 33' (10 mils) E
SW corner	SE corner
1° 02' (18 mils) E	0° 32' (10 mils) E

To plot the average direction of magnetic north join the point circled on the south edge of the sheet to the point on the protractor scale on the north edge at the angle estimated for the current year.

True North / Grid North / Magnetic North — Diagrammatic only

Magnetic north varies with place and time. The direction for the centre of the sheet is estimated at 3° 19' (59 mils) west of grid north for July 2004. Annual change is about 13' (4 mils) east.

Magnetic data supplied by the British Geological Survey.

Base map constructed on Transverse Mercator Projection, Airy Spheroid, OSGB (1936) Datum. Vertical datum mean sea level (Newlyn).

HOW TO GIVE A NATIONAL GRID REFERENCE TO NEAREST 100 METRES

SAMPLE POINT: **Goodcroft**

1. Read letters identifying 100 000 metre square in which the point lies .... NY

2. FIRST QUOTE EASTINGS
Locate first VERTICAL grid line to LEFT of point and read LARGE figures labelling the line either in the top or bottom margin or on the line itself 53
Estimate tenths from grid line to point 4

3. AND THEN QUOTE NORTHINGS
Locate first HORIZONTAL grid line BELOW point and read LARGE figures labelling the line either in the left or right margin or on the line itself 16
Estimate tenths from grid line to point 1

SAMPLE REFERENCE NY 534 161

For local referencing grid letters may be omitted

IGNORE the SMALLER figures of the grid number at the corner of the map. These are for finding the full coordinates. Use ONLY the LARGER figure of the grid number. EXAMPLE: 3 1 **7** 000m

INCIDENCE OF ADJOINING SHEETS

The red figures give the grid values of the adjoining sheet edges. The blue letters identify the 100 000 metre square.

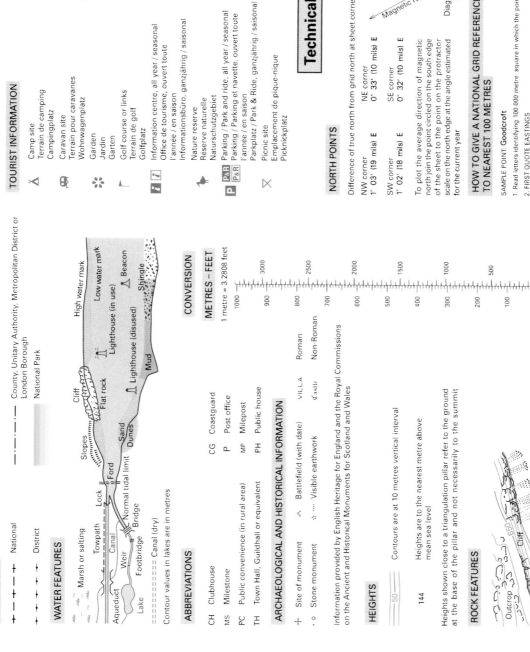

BOUNDARIES

Administrative boundaries as at January 2002

—+—+— National — County, Unitary Authority, Metropolitan District or London Borough
—+—+— District — National Park

WATER FEATURES

Marsh or salting
Aqueduct / Towpath / Lock / Canal / Weir / Footbridge / Normal tidal limit / Bridge / Ford
Lake
- - - - - Canal (dry)

Contour values in lakes are in metres

ABBREVIATIONS

CH	Clubhouse	CG	Coastguard
MS	Milestone	P	Post office
PC	Public convenience (in rural area)	MP	Milepost
TH	Town Hall, Guildhall or equivalent	PH	Public house

ARCHAEOLOGICAL AND HISTORICAL INFORMATION

+ Site of monument ✕ Battlefield (with date) VILLA Roman
· ○ Stone monument ∴ ·· Visible earthwork Castle Non-Roman

Information provided by English Heritage for England and the Royal Commissions on the Ancient and Historical Monuments for Scotland and Wales

HEIGHTS

—50— Contours are at 10 metres vertical interval

144 Heights are to the nearest metre above mean sea level

Heights shown close to a triangulation pillar refer to the ground at the base of the pillar and not necessarily to the summit

ROCK FEATURES

(Outcrop, Cliff, Scree)

CONVERSION
METRES – FEET

1 metre = 3.2808 feet

15.24 metres = 50 feet

Glossary

A

ACCESS how you get to a place

AERIAL PHOTO a photo taken from the air. It can be taken at an **oblique** angle (slanted) or a **vertical** angle (bird's eye view)

ATLAS book of maps

C

CHANNEL the space between two river banks, where a river flows

CLIMATE the average pattern of weather over many years

CLIMATE CHANGE any long-term change in weather patterns

COMPARISON GOODS high-cost goods, like clothes or furniture, that we like to compare before buying

CONDENSATION the change from gas to liquid

CONSTITUENCY a small area of the UK that elects its own Member of Parliament (MP)

CONTOUR LINE a line on a map that joins places at the same height above sea level

CONVENIENCE GOODS low-cost goods, like milk or newspapers, that we buy somewhere that is convenient

COUNTY a smaller area of the UK with its own local government

D

DEMOCRACY a system of government based on people voting

E

ECONOMIC ACTIVITY work that people are paid to do

ECONOMIC MIGRANT someone who moves to another country for work

ECOSYSTEM a community of plants and animals and the environment in which they live

ENVIRONMENT our surroundings. It can include both the **natural** environment and **human** environment

ENVIRONMENT AGENCY the organisation that looks after the environment in England and Wales, including rivers

EVACUATE move people from an area of danger

EVAPORATION the change from liquid to gas

F

FLOOD an overflow of water from a river or the sea

FLOODPLAIN the flat area around a river that floods easily

FUNCTION purpose or use, for example, one function of a town could be as a market

G

GRID REFERENCE the location of a place on a map using the numbers on the grid lines. A **four-figure grid reference** gives the grid square while a **six-figure grid reference** gives the location within the square

GROUNDWATER water that sinks into the ground

H

HIGH POPULATION DENSITY a crowded area

HYPOTHESIS a theory, or idea, that can be investigated to prove if it is right or wrong

I

IMMIGRANT a person who comes to live in a country

IMPERMEABLE does not let water pass through

INCINERATOR a large building where rubbish is burnt

INCLUSIVE included with others

INDUSTRIAL CITY a city that grew up around factories where people came to work

INDUSTRIAL REVOLUTION the period of history from around 1750 to 1900 when many factories were built in Britain

Glossary

INVADER someone who enters a country to take over by force

L
LANDFILL SITE a large hole in the ground in which rubbish is buried

LOW POPULATION DENSITY empty, or few people in an area

M
MARKET TOWN a town that grew up around a market where people came to buy and sell

MULTICULTURAL a community made up of people from different cultures

P
PHYSICAL GEOGRAPHY the study of planet Earth, including natural landscapes, rivers, weather and climate

PLAN view of an object or place from above

POLLUTION contamination of air or water by chemicals, usually from human activities like industry

POPULATION DENSITY the average number of people per square kilometre

PORT a settlement that grew up around a harbour on a river or the coast

PRECIPITATION rain, sleet, hail or snow

PREVAILING WIND the most common wind direction in a particular place

PRIMARY ACTIVITY the way that people obtain a raw material from the land or sea

R
RECYCLING re-using resources over and over to cut down on waste

REFUGEE someone who is forced to flee from danger in their own country

REGENERATION renewal of a declining area – urban regeneration is the renewal of a declining area in a city

REGION a large area of the country with its own geographical characteristics

RELIEF the shape of the land

RESOURCES things that we get from the Earth that we use, for example, water, food, coal

RIVER EMBANKMENT a river bank that is raised to prevent the river from overflowing

RUN OFF the movement of rainwater over the land, including rivers

RURAL AREA countryside, where people live on farms and in villages

S
SCALE the ratio of the distance on a map to the distance on the ground, for example, 1:10

SCALE PLAN a plan drawn according to the real size

SECONDARY ACTIVITY manufacturing, or making a product from raw materials

SEGREGATED separated from others

SETTLEMENT any place where people live, from a single house to the largest city

SITE the land a settlement is built on

SPRING the point where water flows out of the ground

SYSTEM a way of understanding how the world, or different parts of it, work

T
TERTIARY ACTIVITY a service that helps people, but does not produce anything visible

TRIBUTARY a river that flows into a larger river

U
URBAN AREA a built-up area, like a town or city

URBANISATION an increase in the proportion of people living in urban areas

W
WATER CYCLE the never-ending circling of water between the land and the air

WATER VAPOUR water in the form of a gas

WEATHER what happens in the atmosphere from day to day

Index

access plans/surveys 13, 14–15
accessibility problems 10–11, 14–15
acid rain 109
aerial photos 22, 50–1, 88
air pollution 108, 112
   harmful gases 109
aspect 42
atlases 28, 50–1
attitudes, and access 16–17
Avon, River 88–9, 92–7

bays 36–7
beaches 36–7
Boothorpe, Leicestershire 107
British Asians 61
British Isles 29, 117
   capital cities 116
building materials 54, 55
built-up areas, and flood risk 90

canals 56
Canary Wharf, London 5, 7
Celtic tribes 54, 118
chemicals 109
   and water pollution 106, 107, 112
Cheshire Plain 55
cities
   high population density 121
   see also Manchester
cliffs 36–7
climate 42, 123
climate change 84
clouds 40
coastal erosion 4
Colorado River, USA 5, 7
comparison (high-value) goods 74, 75
compass rose 21
concept maps 85–98
condensation 40
connections 20–33
   global 20, 30–1, 33
   local 20, 26–7, 32
   national 20, 28–9, 33
consumers 44, 45
contour lines 38, 39, 50–1
convenience goods 74
cotton 31
cotton mills 56, 60, 62

democracy in action 128–9
disability and education 10–19
   wheelchairs in school 11–16, 18–19

distance, on maps 25, 50–1
earthquakes, causing tsunamis 49
economic activities 114
   primary, secondary and tertiary 126–7
economic immigrants 119
ecosystems 44–5
education 127
electricity, from burning waste 108, 112
embankments 91
emergency services 85–6, 96–7
environment, and waste 100
Environment Agency 87, 91, 95–6, 99
environmental quality evaluation charts 124–5
evaporation 41

factory work 126
farming 126
fertile land 54, 55
Finland 20, 30
flood damage 86–7
flood defences 90
flood graph 94
flood warnings 87, 91, 95
   Severe Flood Warning 95
floodplains 88, 91
floods/flooding
   causes, effects and responses 84–99
   factors increasing risk 90
   factors reducing risk 90, 91
   family flood plan 87
food 44–5, 115
forest 36–7, 45
four-figure grid references 27
fresh water/drinking water 40–1
fuel supply 54

geography 6
   personal 8–19
   saved by geography lessons 48–9
geography passports 132–3
Germany, recycling of waste 105
global warming 106
government in the UK 128–9
   local government 129
   parliamentary constituencies 129
   regional government 128
graphs see flood graph; line graphs; living graphs

Great Britain 117
grid references 27, 50–1, 88
grocer's shop, traditional 72–3
groundwater 40

health 62, 107, 109
height, and temperature 42
homes as systems, inputs and outputs 102–3
household waste see waste
housing 71
   see also regeneration
hydrograph, River Avon 94–5

immigrants 60, 62, 118, 119
immigration 60–1
   into the UK 118–19, 130–1
Imperial War Museum 66
impermeable rock 90
incineration of waste 104
   advantages/disadvantages 108–9
   atmospheric pollution 108, 109
   locations for 109
India 30
Indonesia 20
industry, in Manchester 56–9
internet 30, 80–1, 82
invaders 119
islands
   British Isles 43
   see also Rig Rha island

jobs see economic activities

keys (maps) 38, 50–1
Kosovo, Albania 4, 7

lakes 36–7, 40
landfill sites 104, 106–7
   advantages/disadvantages 106
   opencast mining pits 106, 107
   background 88–9
line graphs 59
living graphs 62–3
London 20, 121, 123
   Wanstead 22–5
low-cost flights 31
Lowry Centre 66, 68
Lowry, Laurence Stephen 58–9

Manchester 20, 52–69
   changing functions 64–5
   city centre regeneration

INDEX

66–7
  East Manchester, Hulme, Sportcity 67
  growth of 56–7
  modern functions 65
  a multicultural city 60–1
  problems, and solutions 66–9
  Trafford Park, Salford Quays 66
Manchester Ship Canal 57, 62
map reading 38–9
maps 23, 24, 26, 29, 64, 74, 89, 116, 122
  concept maps 85–98
  distance measuring 25, 50–1, 88
  Ordnance Survey (OS) symbols 132
  population density 120
  see also atlases; reference maps
marshland 36–7
methane gas 106, 107
MetroLink, tram system 67
mills, steam-powered 62
moorland 36–7
moorland ecosystems 45
multiculturalism 60–1

Normans 118
Newquay 26

Ogden Bridge 71–83
Ordnance Survey (OS) symbols 132

Pennines 55, 116
Phuket, Thailand, tsunami 48–9
pie charts 78–9
plans, drawn to scale 14–15, 19, 50
pollution 57, 62, 86
  atmospheric 108, 109, 112
  from landfill sites 106, 107, 112
population density 120–1
precipitation 40
producers 44, 45

railways 57, 63
rainfall 42, 90
  average annual, UK 123
  and river level 94–5
recycling of waste 104, 105
  re-uses resources 110–11
  WyeCycle 110–11
reference maps 134
refugees 4, 114–15, 119
  see also immigrants
regeneration, Manchester 65,

66–7, 68–9
relief
  contour lines 38, 39, 50–1
  and population density 120, 121
Rig Rha island 34–52
rivers 40
  erosion 5
  and flooding 90–1
rockpool ecosystems 44
rockpools 36–7, 45
Romans 54–5, 118
routes, following/describing 25, 50–1
rubbish see waste
run off 40
rural environment 124

Salford Quays 66
Saxons 118
scale
  of maps 25, 88
  scale plans 14–15, 19, 50–1
Scarborough, Yorkshire 4
schools 8–19
  segregated or inclusive 18–19
Scotland, low population density 121
sea 41, 45
secondary schools
  with wheelchair access 12–19
settlements 88
  best sites for 54
Severn, River 88–9, 92–3, 96–7
shelter 42–3, 54
shopping 70–83
  changes to shopping habits 82
  connections through 31
  high street 75
  Ogden Bridge, shop closures 72
  on-line 80–1, 82
  out-of-town centres 75, 82
  village store 72, 74
shopping surveys 76–9
six-figure grid references 27, 39
sketch maps 55, 117
sketches, labelled 23, 50–1, 88
spider diagrams 9
springs 36–7, 40
steep slopes, and flooding 90
streams 36–7, 55
street maps 25
Sun 42, 44
superstores 71, 82
  compared with village stores 72–3

out-of-town 75
survival 34–51
symbols 27, 38, 50–1, 134

temperature 42, 123
Tewkesbury, Glos, flooding 84–6, 87–8, 92–3, 96–7
timelines 56–7, 119
tornado 86
tourism, in Manchester 65
trade 56
traffic, increase in 71
Trafford Park 66
transport 54, 56, 57, 62
  to landfill sites 106
trees, planted near rivers 91
tributaries, increase flood risk 90
tsunamis 48–9

United Kingdom, and immigration 114–31
urban environment 125
urbanisation 59
USA 5, 20, 30

valleys 36–7
Vikings 118
village shops 71, 72–3, 74

Wanstead, London 22–3
  getting to school 24–5
  leisure activities 26–7
waste 100–13
  excuses for not recycling 111
  organic and non-organic 110, 112
  types of 101, 104–5
  the WyeCycle 110–11
water cycle 40–1
water supply 54
water vapour 41, 42
waterfalls 36–7
weather 42
  in the UK/British Isles 43, 115, 122–3
wind direction, prevailing wind 42

139

→ Acknowledgements

Author's acknowledgement

I would like to thank Liz Taylor, Josie Wood and Vicki Haynes for their helpful ideas and comments. A special thank you to Nancy for being the star of Unit 1. Also, thanks to Ann Tanner and the staff at the Environment Agency in Sussex for their advice on Unit 6. Last, but not least, thank you to Joe and Matt for doing the activities and being my most honest critics!

Photos
Cover *(main photos)* *l* David Parker/Science Photo Library, *r* Robert Harding/Digital Vision, *(inset photos)* *t* Vicki Coombs/Ecoscene, *c* © Brauchli David/Corbis Sygma, *b* F. Sierakowski/Rex Features; **p.4 and 7** *t* John Giles/PA/Empics, *b* © Brauchli David/Corbis Sygma; **p.5 and 7** *t* © Owaki – Kulla/Corbis, *b* London Aerial Photo Library; **p.8** Andy McGuire; **p.9** Andy McGuire; **p.10** *t* Andy McGuire, *b* plainpicture GmbH & Co. KG/Alamy; **p.11** *t* © John Walmsley/ Education Photos, *bl & br* Last Resort Picture Library; **p.12** Andy McGuire; **p.14** Andy McGuire; **p.16** Andy McGuire; **p.18** Andy McGuire; **p.19** © Crown copyright material is reproduced with the permission of the controller of HMSO and Queen's Printer for Scotland; **p.20 and 30** © Prodeepta Das; **p.22** © BLOM Aerofilms; **p.23** © Environment Agency copyright and/or database right 2008. All rights reserved; **p.26** © Philippe Devanne - Fotolia.com; **p.28** *t* Robert Judges/Rex Features, *c* Paul Thompson Images/Alamy, *b* Robert Harding Picture Library Ltd/Alamy; **p.48** Sipa Press/Rex Features; **p.49** Gemunu Amarasinghe/AP/Empics; **p.52** *t* © Aidan O'Rourke, *cl* Ilpo Musto/Rex Features, *bl* Paul Thompson Images/Alamy, *br* Sefton Photo Library/Rex Features; **p.53** *tl* Len Grant Photography/Alamy, *tr* David Crausby/Alamy, *c* Fotofacade/Alamy, *b* Len Grant Photography/Alamy; **pp.58–9** The Lowry Collection, Salford; **p.60** Photofusion Picture Library/Alamy; **p.61** Roy Hsu/Alamy; **p.63** *l* The Lowry Collection, Salford, *r* Photofusion Picture Library/Alamy; **p.70** Peter Titmuss/Alamy; **p.73** *t* Mary Evans/Bruce Castle Museum, *b* TopFoto; **p.74** Ashley Cooper/Collections; **p.75** *t* Nigel Hawkins/Collections, *b* Ashley Cooper/Collections; **p.76** David R. Frazier Photolibrary, Inc./Alamy; **p.81** image from Tesco.co.uk; **p.84** Daniel Berehulak/Getty Images **p.86** Matt Cardy/Getty Images **p.87** Rex Features; **p.88** Matt Cardy/Getty Images; **p.91** © Environment Agency copyright and/or database right 2008. All rights reserved; **p.92-3** Stephen Hird/Reuters/Corbis **p.97** *t* David jones/PA Archive/ PA Photos *b* Leon Neal/AFP/Getty Images; **p.99** reproduced by permission of the Environment Agency; **pp.100–01** Jim Belben; **pp.102–03** Chris Howes/Wild Places Photography/ Alamy; **p.105** Kevin Foy/Alamy; **p.106** Robert Brook/ Science Photo Library; **p.108** Robert Brook/Science Photo Library; **p.110** *all* John Widdowson; **p.111** Photofusion Picture Library/Alamy; **p.113** *l* Robert Brook/Science Photo Library, *c* John Widdowson, *r* Robert Brook/Science Photo Library; **pp.118** Janine Wiedel Photolibrary/Alamy; **p.119** *tl* TopFoto, *tc* Getty Images, *tr* © H.Davies/Exile Images; **p.121** *t* John Widdowson, *b* Andrew Holt/Alamy; **p.124** *t* Photofusion Picture Library/Alamy, *b* Travel-Shots/Alamy; **p.125** *l* Andrew Holt/Alamy, *r* © Molly Cooper/Photofusion; **p.126** *l* Mary Evans Picture Library, *r* TopFoto; **p.127** © John Walmsley/ Education Photos; **p.131** Last Resort Picture Library.

t = top, *b* = bottom, *l* = left, *r* = right, *c* = centre

Text
p.48 extract from article on Tilly Smith from The *Sun* newspaper, January 2005.

Maps and diagrams
pp.24 and 25 (extract) map of Wanstead Street Atlas of London © 2003 Philip's © Crown copyright 2003; **pp.26 and 27** Reproduced by kind permission of Ordnance Survey on behalf of HMSO. © Crown copyright 2007. All rights reserved. Ordnance Survey Licence number 100036470i ; **pp.56 and 62** map of Manchester in 1650, Manchester Library and Information Service: Manchester Archives & Local Studies; **pp.57 and 62** map of Manchester in 1850, Ordnance Survey; **p.64** Manchester city centre street map 1:17 500 © Automobile Association Developments Ltd 2005 A02835. All rights reserved and includes mapping data supplied by © Ordnance Survey, Crown Copyright 2005. All rights reserved. Licence 399221; **p.89** Tewkesbury and the River Severn 1:50 000, Ordnance Survey; **p.94** redrawn hydrograph from flood report July 2007. Copyright © Environment Agency; **p.95** flood watch symbols. Copyright © Environment Agency; **p.133** map of the UK © 2003 Philip's; **pp134–35** Ordnance Survey map symbols.

All OS material reproduced by permission of Ordnance Survey on behalf of HMSO. © Crown Copyright [2003]. All rights reserved. Ordnance Survey number 100036470.

Every effort has been made to contact copyright holders, but if any have been inadvertently overlooked the publishers will be pleased to make the necessary arrangements at the earliest opportunity.

→ Where next?

These are the kind of questions you will be studying in *This is Geography* Books 2 and 3.

Book 2

- **Are you an optimist or a pessimist – do you think Earth Village is heading for disaster?**
- **How will you persuade your teacher to let you spend more of your geography lessons studying football?**
- **Changing fumes! How would you redesign your friend's house to make it more energy efficient?**
- **Plan an adventure trail through a limestone landscape – can you make rocks really interesting?**

Book 3

- **Journalists needed! Write a feature article for a newspaper about the Indian Ocean tsunami.**
- **Burgers and chips on the menu again! How can you improve eating habits in your school?**
- **Be inspired by London 2012 – what will you put in your hometown's bid for the Olympics?**
- **How will you persuade a meeting of world leaders that you have the best plan for the future of Antarctica?**